Random Numbers Unveiled

"Who would have thought a book on random numbers would be more than randomly interesting? Author Szpiro draws readers in with clarity, wit and readily understood examples. Plenty of history, examples and some solid math for the cognoscenti!"

—**Vint Cerf**, *Internet Pioneer*

"George Szpiro, has managed to explain, in his inimitable, gripping and engaging style, with bite-size chapters, the many facets of randomness. I learned so much from this little book!"

—**Prof. Doron Zeilberger**, *Rutgers University*

Random numbers are immensely important in scientific research, in economic decision-making, in polling, gaming, and cryptography and algorithm design. Surprisingly, while many popular books in mathematics have been written about prime numbers, about π, e, $\sqrt{-1}$, there exist no books for the general reader about random numbers. True, the subject of randomness as such has been the subject of several books, but random numbers are only mentioned, if at all, as a by-product. Given the immense theoretical and practical importance of random numbers, this is astonishing.

This book proposes to fill that gap.

The book discusses random numbers under five headings: What are they? What are they good for? How do we produce them? Why do we need them? How do we fake them? The book has been written with a sophisticated general reader in mind, but should be of much interest to students and academics of all levels who have an interest in mathematics and randomness.

Features

- Written in an easily readable, conversational style
- Aimed at general readers who are interested in mathematics in general, or who have read books about π, e, $\sqrt{-1}$, or irrational numbers
- Accessible to anybody with a high school mathematics education

Random Numbers Unveiled

The Secrets of Numbers That You Can't Predict but Can Rely On

George G. Szpiro

CRC Press
Taylor & Francis Group
Boca Raton London New York

CRC Press is an imprint of the
Taylor & Francis Group, an **informa** business

A CHAPMAN & HALL BOOK

Designed cover image: David Amir

First edition published 2026
by CRC Press
2385 NW Executive Center Drive, Suite 320, Boca Raton FL 33431

and by CRC Press
4 Park Square, Milton Park, Abingdon, Oxon, OX14 4RN

CRC Press is an imprint of Taylor & Francis Group, LLC

© 2026 George G. Szpiro

ISBN: 978-1-041-07644-5 (hbk)
ISBN: 978-1-041-07329-1 (pbk)
ISBN: 978-1-003-64152-0 (ebk)

DOI: 10.1201/9781003641520

Typeset in Minion
by Newgen Publishing UK

*In memory of my mother, Marta Szpiro
(1927, Beregszász–2025, Zürich), a Holocaust survivor
who—together with my late father, Simcha Binem—built a
family and a life of love, dignity, resilience, and humor.*

Contents

PART III **Random Numbers: How Do We Produce Them?**

PART IV **Random Numbers: Why Do We Need Them?**

PART V   **Random Numbers: How Do We Fake Them?**

Foreword

I first came across the writing of George G. Szpiro more than twenty years ago when I wrote a rave review, in the journal Science, of his 2003 book on Kepler's conjecture, that I really loved. This was followed by fascinating books on the Poincaré Conjecture, the economics of risk, and most recently a lovely compendium of paradoxes. I also loved his Mathematical Medley, a collection of short mathematical essays that won a media award. In particular, one of the essays was a delightful portrait of my beloved silicon disciple, Shalosh B. Ekhad.

Mathematicians are often accused—rightly—of taking themselves too seriously. We polish our definitions, prove our theorems with rigor bordering on obsession, and shun anything that smells of uncertainty. And yet, paradoxically, nestled at the heart of our most precise theories lies the imp of chaos: randomness and its incarnation as random numbers.

Random numbers are both tricksters and saviors of modern mathematics. We recoil at their unpredictability while relying on them to power simulations, fuel cryptographic protocols, and even, curiously enough, help prove theorems. When deterministic strategies fail us, we turn to random numbers like gamblers at a roulette table…only we dignify this by calling it the Monte Carlo method.

The author of this volume has taken on the delightful task of wrestling with this paradox. *Random Numbers Unveiled* is much more than yet-another technical manual. It is a wide-ranging exploration of randomness and random numbers in their many roles: as philosophical puzzle, mathematical tool, cultural force, and computational necessity. The author takes us on a guided journey through probability, epistemology, computation, and cultural history. He asks: What does it mean for a number to be random? Can humans generate randomness? Should we trust machines that do? And most daringly do random numbers actually exist, or are they just figments of our mathematical convenience? From parapsychology to pseudo-random generators, from sortition in Athenian democracy to zero-knowledge proofs, the author traces how our species has coped with, celebrated, and sometimes misunderstood randomness.

With wit, clarity, and a critical eye, this book invites the reader to think deeply-and sometimes irreverently-about how we quantify uncertainty and whether doing so reveals truths about the world, or merely our limitations in knowing it. What you will find is something very rare: curiosity sharpened into insight, and a keen appreciation for the profound absurdities that lie beneath the surface of seemingly simple concepts.

If you're a mathematician, scientist, or just a curious reader with a high tolerance for epistemological dizziness, you'll find this book both illuminating and fun. And if, after reading it, you're still not sure what random numbers *really* are—well, that's part of the point.

In any case, you are in for a treat. Read on. And may your journey through randomness be delightfully unpredictable.

Doron Zeilberger
Department of Mathematics
Rutgers University

Preface

THIS BOOK IS ABOUT the curious, elusive, and indispensable nature of random numbers.

When I was a student of mathematics in the late 1960s at the Swiss Federal Institute of Technology (ETH), the curriculum was all about strict definitions, the precision of arguments, the clarity of structure, the rigor of proofs. The elegance of analysis, algebra, geometry, topology, and the absolute certainty that axiomatic, no-nonsense reasoning provides, seemed diametrically opposed to the apparent messiness of these curious thingies that were being conjured up by some new-fangled computing machines: random numbers.

Computer science was then an emerging field and no more than a subdivision in the department of mathematics – albeit with some important forerunners of the field on the faculty. But haughty students that many of us were, we reveled in the beauty of pure mathematics, scorning anything applied. 'Data processing,' as the field was then called at the ETH, was for nerds and technicians.

Tempora mutantur, nos et mutamur in illis (The times are changing, and we change along with them). Even the haughty, if naïve, idealists of yore now realize that random numbers are all around us: in the shuffle of cards, in the timing of radioactive decay, in the algorithms that secure our digital communications, in the Monte Carlo methods that let us simulate the complex world we inhabit, etc. And random numbers have a beauty of their own. Though we seldom pause to consider what they truly are, what they are good for, how we produce them, why we need them ... and how we can fake them ... we now find ourselves relying on them daily.

In the pages that follow, we explore the uneasy coexistence of order and randomness. We will meet the mathematicians, computer scientists, physicists, cryptographers, and philosophers who have grappled with the embodiment of randomness in random numbers, and examine the surprising ways these numbers shape our world. Along the way, we will delve into the paradoxes and practical challenges that arise when trying to pin down something that, by its very nature, resists being pinned down.

୨

The book is meant for the general reader with an interest in mathematics, computer science, and the hidden workings of the modern world, as well as for students and professionals who may wish to deepen their understanding of random numbers, both as a concept and as a tool. While some chapters contain equations and technical descriptions – mostly this doesn't go much beyond a relatively low-level understanding of high-school algebra – they are accompanied by intuitive explanations, stories, and examples, so that readers can enjoy the book even if they choose to skim the more technical sections.

Random numbers are everywhere; and they matter! I hope this book will help you see why.

Author Biography

George G. Szpiro, born in Vienna, holds dual Swiss and Israeli citizenship. He earned his MSc in mathematics from the Swiss Institute of Technology (ETH) in Zurich, an MBA from Stanford University, and a doctorate in mathematical economics from the Hebrew University in Jerusalem. Following consulting stints with McKinsey and Company, he turned to academia. After several years of academic research and teaching at the Wharton School (University of Pennsylvania) and the Hebrew University, he changed his career again and became a foreign correspondent and mathematics columnist for the Swiss daily *Neue Zürcher Zeitung*. For over 20 years he reported from Israel, and for 6 years from New York City.

Apart from two dozen academic papers in mathematics, statistics, economics, and genetic algorithms, Szpiro has authored nine books for general-interest readers in mathematics, economics, and political science. He has been awarded the *Prix Média* by the Swiss Academy of Natural Sciences, the *Medienpreis* of the German Mathematical Society, and was a finalist for the *Descartes Prize* of the European Union.

Books by George G. Szpiro

- *Kepler's Conjecture: How Some of the Greatest Minds in History Helped Solve One of the Oldest Math Problems in the World* — John Wiley & Sons, 2003
- *The Secret Life of Numbers: 50 Easy Pieces on How Mathematicians Work and Think* — Joseph Henry Press (National Academy of Sciences), 2006
- *Poincaré's Prize: The Hundred-Year Quest to Solve One of Math's Greatest Puzzles* — Dutton (Penguin Group), 2007

- *Numbers Rule: The Vexing Mathematics of Democracy, from Plato to the Present* — Princeton University Press, 2010
- *A Mathematical Medley: Fifty Easy Pieces on Mathematics* — American Mathematical Society (MAA Press), 2010
- *Pricing the Future: Finance, Physics, and the 300-Year Journey to the Black-Scholes Equation* — Basic Books, 2011
- *Risk, Choice, and Uncertainty: Three Centuries of Economic Decision-Making* — Columbia University Press, 2020
- *Perplexing Paradoxes: Unraveling Enigmas in the World Around Us* — Columbia University Press, 2024
- *Ignorance: What We Do Not Know, Cannot Know, Must Not Know, and Refuse to Know* — Columbia University Press, 2026

Introduction

RANDOM NUMBERS NEED COMPANY

In December 1969, with the Vietnam War raging, the US Military required additional recruits. The Selective Service System, the US agency which maintains the database for the recruitment of military personnel, thought they had found a selection process that would be fair: select recruits randomly by their birthdays. Slips of paper with the 366 dates of the year (including February 29) were placed in plastic capsules which were then placed in a glass jar. Random numbers were drawn, one by one, with the first 195 dates corresponding to the birthdates of young men who would probably be drafted. The lower the draft number, the higher the likelihood that the young man would go to war. By relying on random numbers drawn from the jar, the process seemed fair enough.

But it turned out that the selection was not random! Of the 31 days in December, 26 were ranked below 195. A ranking of the months by the number of birthdays which would probably be spared the draft revealed that the months January to June were all ranked above the months July–December, truly a source of frustration for young men born in the second half of the year, especially those born in December.

DOI: 10.1201/9781003641520-1

FIGURE   Congressman Alexander Pirnie (R-NY) drawing the first capsule for the Selective Service draft, December 1, 1969.

https://commons.wikimedia.org/wiki/File:1969_draft_lottery_photo.jpg

What had happened? Journalists surmised the following: the capsules with the January dates were placed into the jar first. When the February dates were put into the box, they were mixed up with the January capsules. The March dates were mixed with the January–February heap, and so on. Altogether the January capsules were mixed 11 times, while the December dates were mixed only once. Somehow the capsules containing the dates of the early months must have remained, or come to lie, towards the bottom of the bowl and those containing dates of the late months towards the top. And the person who picked the capsules may have picked more from the top than from the bottom. The dates were not drawn at random! And by assigning them sequential numbers – 1, 2, 3, …, 366 – the process was not fair.

To make future draftings of recruits more random and equitable, the Selective Service System took a page from the Jewish Talmud. When Elazar, the High Priest, needed to divide the Holy Land among the 12 tribes, he used not one urn for randomization, but two:

> An urn with the names of the twelve tribes, and an urn containing descriptions of the boundaries were placed before him. … he shook the urn of the tribes and the name Zevulun came up in his

hand, he shook the urn of the boundaries and the boundary lines of Acco came up in his hand. Then he shook the urn of the tribes and the name Naphtali came up in his hand. He, shook the box of the boundaries, and the boundary lines of Gennesar came up in his hand. And so it was with every tribe.

In subsequent draft lotteries, the Selective Service System used two drums, one containing the dates, the other containing numbers between 1 and 366. After rotating the drums for an hour, the drawing began. A date was drawn randomly from one drum, say September 16, and a random number from the second drum, say 139. Hence, boys born on September 16 were assigned the draft number 139. By picking not only a date at random but also a random number, the selection process had become fair!

❧

Random phenomena have always fascinated humanity. Since the earliest ages of recorded history, humans have tossed knuckle bones, read animal entrails, and interpreted signs in the natural world on the belief that these random phenomena conveyed the will of the Gods. Only in the last few centuries have we truly come to appreciate that behind the myth and ritual there has been an astonishing and baffling truth: randomness can actually be a source of extraordinary power and insight. This leap, from superstition and awe in the face of randomness to control of its potential, became possible once randomness was harnessed mathematically, in the form of random numbers.

Somewhat paradoxically, random numbers are the medium which allows us to solve many important problems, using disorder itself as a tool to reveal hidden order in our universe. Every day, mathematicians, engineers, decision-makers, among countless others, utilize random numbers to find answers to their questions. Much innovation today is driven by oceans of random numbers – torrents of digits (zeroes to nines) or bits (ones and zeroes) – generated by a mind-bending variety of methods, many of which we will explore in this book.

Though random numbers are of utmost importance in the natural sciences, medical research, social sciences, economics, computer science, gambling, they do not have a good press. Unlike prime numbers, about which many books have been written, hardly any popular accounts of random numbers exist. This is all the more surprising since random numbers are a valuable resource for which interested parties are even prepared to pay good money. It is about time that random numbers received the appreciation, and understanding, they deserve.

ๆ

Unlike odd and even numbers or prime numbers, random numbers do not exist in isolation. The question whether a number like 17 or 14,156 is random is not legitimate: for a number to be random, it needs company.[1] All on its own, a random number does not exist; it obtains the moniker 'random' only when it is part of a collection of random numbers.

If this sounds a bit like a tautology, it is. And that is not the only paradox. Because even when it is surrounded by many of its ilk, it is by no means easy to determine whether a sequence of numbers is random. Though we can identify many sequences that are *not* random simply by inspecting them, the only acceptable manner of certifying randomness is to study how the sequence came about...and even then, randomness is not guaranteed.

Hence, randomness is not a characteristic of a number itself but of a collection of numbers. A collection of uniformly distributed numbers is said to be random, if the numbers appear approximately equally often, and each one is independent of the previous ones. A sequence of numbers is truly random if it fulfills the sacred trinity of randomness – the numbers must be *u*npredictable, *u*niformly distributed, and *u*ncorrelated (U³).

For example, the numbers zero to nine are uniformly distributed if each digit appears approximately 10% of the times. Binary digits, that is, zeroes and ones, are uniformly distributed if each appears approximately half the times. There are other distributions, for example, the normal distribution which follows the so-called bell curve. But in this book, unless otherwise noted, I only speak about uniformly distributed numbers. In all cases, however, for a sequence of numbers to be random, one must not be able to compute or predict or guess any of the numbers that follow.

So let us see how random numbers are generated. In contrast to even numbers that can be generated by taking any number whatsoever and doubling it, or odd numbers that can be created by taking an even number and adding 1, or transcendental numbers like π, √2, or *e* (Euler's number), that can be computed by algorithms, the generation of random numbers presents immense difficulties.

---

[1] Even a single coin throw that determines which side kicks the ball first in a soccer game, or who gets the white pieces in a chess game, can be considered as random only if it is seen as one throw in a long series of throws.

In general, one must resort to physical devices: coins, dice, floating ping-pong balls, roulette balls, and the like. However, such tools are slow. For the reams of random numbers that are required nowadays for even moderately large projects, tossing coins or rolling dice would be far too sluggish. And anyway, who can guarantee that a coin or a die is unbiased?

Just like most difficulties nowadays are solved by letting computers deal with them, one tries to do the same with the vexing problem of generating random numbers. But that creates a host of new problems. Computers are machines that are deterministic, that is, the exact opposite of random! Hence, as we shall see presently, computers can at best only generate *pseudo*-random number sequences, that is, sequences of numbers that *seem* random, though, strictly speaking, they are deterministic.

So, as a last recourse, if you don't trust deterministic algorithms to generate random sequences of numbers, it's back to using tools, albeit more sophisticated ones than coins and dice: atmospheric noise, the movement of computer mice, lava lamps…and the *non plus ultra*, quantum effects.

THE BOOK AT A GLANCE

In Part I of the book, I discuss what random numbers actually are, and then describe early efforts to generate sequences of random numbers and various methods to test them. For that, we must ask ourselves a fundamental question: do random number sequences actually exist? More generally, is randomness real? Or is everything in our world deterministically predetermined and the only reason why we perceive something as random is because we are ignorant of some underlying causes and reasons? For the faithful the answer is simple: it is the finger of God that determines everything; for agnostic rationalists, the jury is still out. And speaking of God, if it is not His finger that generates random number, the question is whether human beings can create, or even just recognize, such sequences. (Spoiler alert: they cannot.)

In order to create random numbers, people have been playing with bones, stones, and dice-like solids since ancient times; more recently, playing cards, roulette wheels, floating ping-pong balls have been used. Depending on the implements, there are good and less good sequences of pseudo-random numbers. Hence, it would be nice to have a measure of their quality. The key notion that is used to quantify the amount of randomness that is present in a sequence of numbers is 'entropy,' a statistical concept of disorder. It was introduced in the second half of the nineteenth century by German, Austrian, Scottish, and American scientists who studied the motion of molecules in thermodynamics, and adapted in the twentieth century to define the information content that a message or a sequence of numbers contains.

Once we know what random numbers are, we turn, in Part II, to the question of what they are good for. Random coin tosses, throws of dice, rolls of roulette balls, distribution of playing cards are the cornerstone of gambling activities. (As may be expected, games of luck attract the curious, the adventurous, and also the unsavory, and I will recount some attempts to dupe, scam, and defraud gambling establishments.) Although gaming and gambling come to mind most frequently, there are more areas of daily life in which randomness is required. Polling and sampling, for example, require random numbers. But as some pollsters have found out to their chagrin, it does not suffice to throw random numbers at a problem. It may sound like an oxymoron, but random polls must be carefully designed; otherwise, serious blunders can, and did, occur.

An entirely different, and unexpected, use for randomness is the replacement of elections and decisions by lotteries. Making decisions by drawing lots, a method called 'sortition,' may not be the silliest scheme since decision by majority vote, the bedrock of democracy, is vastly overrated. Even in the financial industry, the main takeaway of many studies is that recommendations from investment advisors are just as good, and just as bad, as coin throws.

Now that we know what random number sequences are good for, let us see, in Part III, how they can be generated. Besides coins and dice, there are other objects that produce randomness: many-sided polyhedra, playing cards, spinning tops. And since the early nineteenth century, many patents have been granted for machinery that is more sophisticated than the throw of a die. Nevertheless, the closest one can hope to get to pure randomness is by resorting to physical phenomena, as, for example, the chaotic movement of wax blobs suspended in water, the crackle of radio static, the noise emanating from computer hardware.

As of today, the ultimate source of random numbers are the random occurrences of quantum phenomena. Unfortunately, however, given the inevitable imprecision of hardware setups, even quantum mechanics does not provide honest-to-goodness, impeccable, true random numbers.

Part IV turns to the realm of computer science. Cryptography is an essential tool of today's life; it permits purchases of goods and services, transmission of private messages over the internet, and much more. However, to protect people's privacy, data and messages must be encrypted which requires prime numbers that are selected randomly. Furthermore, for certain encryption methods, numbers are required that are not necessarily prime but that are necessarily random.

By far the most widespread applications of random number is the so-called Monte Carlo method. Originally developed during World War II's Manhattan Project, Monte Carlo method was used by scientists to simulate models of nuclear phenomena rather than bringing atomic bombs to explosion. Nowadays, large-scale models of climate change, scientific phenomena, mathematical problems, aircraft design, business decisions, and economic policies are routinely simulated using Monte Carlo techniques. But even the best-designed generator of pseudo-random numbers can go awry, and there is a history of notorious blunders.

Another cryptographic scheme that employs random numbers is so-called zero-knowledge proofs: by using random bits (binary digits), one party, traditionally called Peggy, the prover, can prove to Victor, the verifier, that she/he knows the answer to a question or the solution to a problem, without actually divulging what it is.

Returning to computer science, there exist many problems that can be solved by algorithms, but only in principle. Often the runtime would increase exponentially as the size of the input increases and even for moderately large inputs, it would take centuries, if not millennia, to arrive at a solution. Sometimes, however, one is in luck and random numbers may come to the rescue like a *deus ex machina* in ancient theater plays. By allowing algorithms to make random choices along the way, runtimes may be sped up massively. As a result, problems that were intractable with deterministic methods can be solved; questions that could not be decided in useful time can be answered quickly.

Unfortunately, the answers sometimes come with a certain probability of being incorrect. However, the secret sauce of such randomized algorithms is that this error probability can be made as small as one likes, until one has, for all practical purposes, reached certainty. But there is a surprise: after extolling the advantages of randomizing algorithms, we make a veritable U-turn and discuss how randomness can be eliminated from these algorithms.

A recurring theme of this book – probably its fundamental takeaway – is the difficulty of generating randomness: random numbers are extremely hard to come by. Part V discusses what can be done about that. As mentioned earlier, computers cannot truly produce random numbers. The best they can do is fake them by running algorithms that generate *pseudo-random* numbers, that is, sequences of numbers that look random, even though they have been generated by a deterministic computer.

Some number sequences that one encounters contain more randomness, some less. Computer scientists have developed several methods that squeeze as much randomness as possible out of low-random sources. One secret is to use mathematical functions that are so complicated that they work only in one direction: using an initial value, the so-called seed, the functions produce one number after another that seem random since one cannot use them to reckon back to the seed.

While writing this book about random numbers, I became more and more fascinated by the breadth and scope of the subject. Who would have thought that random numbers present such a fertile and rewarding field, not only in mathematics and computer science but also in the practical, day-to-day world? The further I delved into the subject, the more facets I discovered…and I am happy that I did!

I would like to thank Doron Zeilberger, Barry Cipra, Vikrant Ashvinkumar, Rachel Lawrence, Yuliy Lobarev, and anonymous referees for correcting errors and making comments.

I

Random Numbers
What Are They?

CHAPTER 1

Without Rhyme or Reason

What Are Random Numbers?

We all know what random numbers are; at least we think we do: they are numbers that appear by chance, without rhyme or reason, without any pattern.

As I mentioned in the introduction, it is not legitimate to ask whether a number like 17 or 14,156 is random because randomness is a characteristic of a collection of numbers, not of the number itself. A collection of numbers is said to be random if the numbers in the collection are drawn from a certain distribution and there is no correlation between any of them; to anticipate, or predict, or guess any of the following number is impossible. (In this book, we mostly consider uniformly distributed sequences of random numbers but there are sequences that follow other distributions, for example, the normal distribution.)

It should be easy to produce a sequence of random numbers, should it not? Just let them spring 'randomly' from your mind? Well, that won't work; the sprung-from-your-mind sequence will surely not be random; willy-nilly, you try to mix up the numbers, to make them seem random... and that's the problem.

Start an experiment by producing a random one thousand zeros and ones (bits or *binary digits*). How often did eight consecutive ones, that is,

DOI: 10.1201/9781003641520-3

11111111, come up in your sequence? Most probably you avoided such a string of bits because it does not appear to be random. However, in truly random sequence of one thousand bits, this sub-sequence should appear on average about four times. Likewise with 00000000, with 11110000, with 01010101, and with any other string.

Or think of a sequence of five decimal digits between zero and nine. Probably you would not have thought of, say, 12345 or 77777 because such sequences do not seem random at all. However, in a sequence of a million random numbers, these exact sequences should each turn up approximately ten times. A number sequence that is truly random would contain 'unlikely' sub-sequences at just the statistically correct rate.

In fact, it is extremely difficult to produce a random sequence of numbers – binary or digital – and this is what this book is about.

ﾂ

The notion of randomness has been used implicitly, though not recognized for what it is, for millennia. Early playing objects that produced results by chance, such as knuckle bones (ankle bones of sheep, buffalo, or other animals), two-sided throw sticks, dice, have been excavated by archeologists in the Middle East, India, China, and elsewhere.

The oldest examples of six-sided dice have been found in excavation sites in today's Iraq, in Pakistan, in China, and in Egyptian tombs. They date back to the second and third millennium BCE. Marked with dots from one to six, just like today's dice, they produced random numbers when thrown on the floor or table. Legend has it that the Trojans used dice – dare we say 'random number generators' – to while away the time during the siege of their city in the twelfth century BCE.

Of course, the ancient Greeks, the Romans, the Indians, and Chinese did not have a concept of randomness. These cultures used such devices to gamble, to foretell the future, and to make decisions, in the belief that it was a divine being that directed the paths of the dice. Only in the sixteenth century CE did the Italian mathematician, physician, astrologer, physicist, biologist Gerolamo Cardano (1501–1576) realize that the throws of dice were random. By the way, since science alone did not provide sufficient income for this polymath, he supported his family by gambling with dice, thereby discovering the basic concepts of probability theory.

In 1890, the British scientist Sir Francis Galton was one of the first scientists to use dice explicitly to produce random numbers, in order "to test the practical value of some theoretical process, it may be of smoothing, or of interpolation, or of obtaining a measure of variability. or of making some particular deduction or inference." Tossing dice is less tedious than shuffling cards, or drawing balls out of a bag, he declared. And while spinning tops and roulette wheels were preferable to cards or balls, nothing in his opinion could beat dice.

> When they are shaken and tossed in a basket, they hurtle so variously against one another and against the ribs of the basket-work that they tumble wildly about, and their positions at the outset afford no perceptible clue to what they will be after even a single good shake and toss.

Interestingly, Galton used dice with a twist: after throwing a die, he rotated it with his eyes closed and then noted which of the face's four edges pointed away from him. Thus, he obtained four possibilities for each of the six faces, altogether 24. (This corresponds to 4.58... bits.)

One of the first publicly available tables of random numbers was published by the statistician L.H.C. Tippett at Cambridge University Press in 1927. It contained 41,600 random digits, arranged in 10,400 groups of four, drawn from a 1925 census report. Several other collections followed, for example, a table constructed by picking 7,500 two-digit numbers from a large table of logarithms.

But one must beware of tables of logarithms; they can really throw randomness hunters off the track. In the late nineteenth century, when logarithmic tables were widely used by scientists and engineers to perform arithmetic calculations (with logarithms, multiplications are reduced to simple additions), the Canadian astronomer Simon Newcomb noticed that the early pages of books containing the logarithms, that is, the tables starting with 1 or 2, were much more smudged than the later pages that contained the tables starting with 8 and 9. With a leap of confidence he surmised that numbers whose leading digits were small, like 1... were much more frequent than those starting with large digits, like 9...

This was extremely baffling, but in the 1930s, the physicist Frank Benford discovered that numerical values for many data sets – population sizes, physical constants, surface areas of rivers, molecular weights, street addresses, electricity bills, stock prices, housing prices – conformed to what would later become known as Benford's Law: the frequency of small leading digits is high and decreases as the leading digits become larger. One explanation for that puzzling phenomenon is that data, for example, the population of a town, must increase by a 100% to grow from 1… to 2…, but only by 50% to grow from 2… to 3…, and only by 12.5% to grow from 8… to 9… . So, as a source for random numbers, the leading digits of any data set must never be used; only the trailing digits may do.[1]

In 1938, Maurice G. Kendall and Bernard Babington Smith, two British statisticians, constructed a table of 100,000 random numbers using an apparatus akin to a 'wheel of fortune,' a spinning disk divided into ten equal sectors, that was "illuminated from time to time by an electric spark or by a flash of a neon lamp, which is of such short duration that the disc appears to be at rest." At each flash, the illuminated number was recorded; it was the first time a machine was used to generate random numbers.

The two statisticians also coined the notion of 'locally random sequences,' that is, sub-sequences extracted from a random number sequence that also 'look' random. This was new; it derived from the observation that a sequence of, say, one thousand consecutive 7s certainly does not look random. However, in a *huuuuuge* sequence of $10^{1,000,000}$ random digits, sub-sequences of a thousand consecutive 7s should appear approximately $10^{999,000}$ times, practically a certainty.[2] If runs of a thousand zeros did *not* occur every once in a while in such a huuuuuge sequence, it would not be considered random. Still, a sub-sequence of many consecutive 7s, even if drawn from a truly random number sequence, would not be called locally random.

Equally important, Kendall and Babington Smith designed a battery of four tests to determine whether a sequence of random numbers may be considered locally (and globally) random. The first is a simple frequency test: each digit should appear approximately equally often. Hence, decimal digits (the numbers between 0 and 9) should each appear approximately 10% of the times, binary digits (zeros and ones), 50% of the times. This test determines whether the digits are uniformly distributed, Benford's leading digits would definitely have failed it. The next three tests measure the independence of successive numbers from their predecessors.

The serial test is an enhancement of the frequency test: each sequence of two digits should appear equally often. With decimal numbers, any pair – 23, 87, 44, and the like – should appear 1% of the times; with binary

---

[1] By the way, Benford's Law can be easily verified on Google: choose a three-digit number and then check the amount of hits you obtain when this number is appended to 1, to 2, to 3, … .

[2] $10^{1,000,000}/10^{1,000} = 10^{999,000}$.

digits, the pairs 10, 01, 11, 00 should appear one quarter of the times. The serial test could be further enhanced to investigate triples, quartets, and so on, but Kendall and Babington Smith turned to something slightly more sophisticated.

Their third test was inspired by the game of poker. The frequencies of five consecutive numbers should correspond to their probabilities in the card game. Using combinatorial methods, they computed the expected frequencies of 'four of a kind,' 'three of a kind,' 'full house,' 'two pairs,' and 'one pair' in poker and demanded that true random number sequence conform to these frequencies. (Obviously, they included five of a kind in their calculations, which has no counterpart in poker, and, of course, they omitted flushes and royal flushes.)[3]

Finally, they subjected the gaps between identical digits to a test. The digit 7, for example, should be followed by another 7 in about one-tenth of the cases. A gap of one position between the two 7s should occur in about nine hundredth of the cases, a gap of two positions in 81 thousandths of cases, and so on.

The business of generating and publishing random numbers culminated in 1955 with a publication by the aptly, though inadvertently, named RAND Corporation in Santa Monica, California. (RAND does not stand for 'random' but is an acronym for Research and Development).

It was in the late 1940s, when the corporation's engineers performed simulations for the Douglas Aircraft Company and the US Air Force. These required large amounts of random numbers and the only available source at that time was the table by Kendall and Babington Smith. Hence, to perform their simulations, that table's 100,000 random digits would have to be used over and over again, with the consequent danger of introducing unwanted correlations. So, the engineers of the RAND Corporation had to think of something new.

Luckily, these were the founding days of electronic computers and the RAND engineers made ample use of this new toy.[4] The generation of

---

[3] This test is comparable to the definition of a random number sequence by Richard von Mises as a sequence in which no sub-sequence can be chosen that would improve the odds at winning a gambling game (see Chapter 4).

[4] Note that it was important to say 'electronic computers,' because 'computers' *tout court* were not machines but people who did computations.

random digits proceeded in two stages, the first involving hardware, and the second software.

In the hardware stage, an electronic apparatus passed pulses to a five-place digital counter, thus producing decimal numbers between zero and 31. Twelve of these were ignored, which left the numbers between 00 and 19. The second digit of these two-digit numbers was retained, one per second, as the sought-after random number. They were punched on 20,000 punch cards, 50 digits to a card. Placed on top of each other, the stack of cards would have reached a height of 3.5 meters.

Unfortunately, there was a snag. Upon inspection of one block of 125,000 digits, it turned out that there were 62,882 odd and only 62,118 even numbers. Though some deviation from the exact middle was to be expected, and the digits passed the other randomness tests suggested by Kendall and Babington Smith with flying colors, the deviation of 382 even and odd numbers from the expected 62,500 disturbed the RAND engineers immensely. They proceeded to stage 2, the software stage.

Feeding the 20,000 cards into an electronic computer, they added each digit on a card to the digit in the same position on the preceding card, *modulo ten*.[5] These one million re-randomized digits were finally the ones that were to be printed. They were reproduced by photo-offset from the 400 pages printed by an IBM printer, 2,500 digits to a page. "Because of the very nature of the tables, it did not seem necessary to proofread every page of the final manuscript to catch random (sic!) errors…," the unnamed authors explain. Nevertheless, all pages were skimmed to spot systematic errors, every twentieth page was proofread, and the sum of the digits of every fortieth page was checked against the sum obtained from the cards.[6]

The introduction to the book provides some pointers and tricks on how to choose the starting position judiciously (read: randomly) since it is very inadvisable to begin experiments or simulations always at the same spot in the table.

> The lines of the digit table are numbered from 00000 to 19999. In any use of the table, one should first find a random starting position. A common procedure for doing this is to open the book to

---

[5] If the sum was smaller than 10, then this was the number that was kept (e.g., 3 + 4 = 7); if it was greater than 10, then 10 was deducted and that digit was kept (7 + 8 = 15, deduct 10, keep 5). We will have more to say about *modular arithmetic* starting in Chapter 12.

[6] As the book's title indicates, RAND also produced 100,000 normal deviates, that is, random digits that are distributed according to the bell shape, but I won't go into this.

```
73735  45963    78134  63873
02965  58303    90708  20025
98859  23851    27965  62394
33666  62570    64775  78428
81666  26440    20422  05720

15838  47174    76866  14330
89793  34378    08730  56522
78155  22466    81978  57323
16381  66207    11698  99314
75002  80827    53867  37797

99982  27601    62686  44711
84543  87442    50033  14021
77757  54043    46176  42391
80871  32792    87989  72248
30500  28220    12444  71840
```

FIGURE 1.1   A sample of 300 digits randomly selected from RAND's book. (From https://commons.wikimedia.org/wiki/File:Random_digits.png)

an unselected page of the digit table and blindly choose a five-digit number; this number with the first digit reduced modulo 2 determines the starting line; the two digits to the right of the initially selected five-digit number are reduced modulo 50 to determine the starting column in the starting line. To guard against the tendency of books to open repeatedly at the same page and the natural tendency of a person to choose a number toward the center of the page: every five-digit number used to determine a starting position should be marked and not used a second time for this purpose.

There are several creative ways of using the tables:

> Ordinarily, the table is read in the same direction as a book is read; however, the size of the table may be effectively increased by varying the direction in which it is read. Thus, one may read columns instead of lines, may read the table backward, may read lines forward but pages from bottom to top, etc.

The book is still available for purchase and the reviews on Amazon.com are hilarious: "Keeps you guessing until the very end."; "The randomness got repetitive."; "They only used 10 random numbers and just kept repeating them."; "If you've seen one random number, you've seen them all."; "I noticed that each digit appeared almost exactly 10% of the time. Does that sound random to you?"; and finally, the perceptive remark: "I'm kinda puzzled by the purpose of this book. Once published, a list of random numbers becomes predictable."

≈

When George Brown, the chief of RAND's Numerical Analysis Department, recounted the history of the efforts, he ended his account with the wish that in the future

> we won't have to build any more random number generators. It was an interesting experiment, it fulfilled a useful purpose… but it may not be asking too much to hope that … some other numerical process will permit us to produce our random numbers as we need them.

'As we need them' is the key clause in Brown's statement. What the engineer was presciently hoping for was an algorithm that would generate random numbers on the fly, at the required speed and in the necessary quantities, whenever needed. Whether and how this would be achieved will be described in later chapters.

≈

In the late 1930s, when Maurice G. Kendall and Bernard Babington Smith puzzled over how to verify randomness of a sequence of numbers, their checklist consisted of four tests. Six decades later, the American computer scientist George Marsaglia published a CD-ROM with a battery of no less than 15 statistical tests, which he named the 'Diehard Tests.' With names like monkey test, parking lot test, birthday spacings test, squeeze test, random spheres test, they assess the quality of a sequence's randomness. Several of them consider consecutive numbers as coordinates of points in two- or three-dimensional space and measure the distribution of the points' distances to neighboring points. In order for the sequence to qualify as random, the distances must follow certain probability distributions.

The Diehard battery was expanded by Robert G. Brown of Duke University to the 'Dieharder Suite,' which he intended to be the 'Swiss army knife of random number test suites.' The suite is intended not to test number sequences themselves but to test *generators* of random numbers. After all, Brown asked, is the number 7 random? If it is generated by a random process, it might be; if it is made up to serve the purpose of some argument, it is not. Therefore, the Dieharder Suite subjects the processes that generate the supposedly random numbers to tests. I will have much to say about generators of random number and of their cousins, generators of *pseudo*-random numbers, in Part V of this book.

For the time being, take note of a seeming paradox: if you encounter the sequence 1111111111, you would consider it as non-random as could be. But if you actually observe a fair coin being thrown and falling heads ten times in a row, the sequence HHHHHHHHHH would have to be considered random because it was generated by a random process.

As of this writing, the set of statistical tests proposed by the National Institute of Standards and Technology (NIST) is the gold standard for the detection of deviations of binary sequences from randomness. The 130-page report lists no less than 15 tests that are included in the accompanying statistical package. They attempt to detect whether patterns of bits occur more frequently than they statistically should in a truly random sequence.

The cumulative sums test, for example, tests if the bits in a binary sequence are unbiased. Imagine a hiker starting at sea level and taking steps up or down: every time she encounters a '1,' she takes a step up; for a '0,' she takes a step down. Over a long enough journey, if the sequence is random, the hiker should wander around sea level, sometimes going up and sometimes going down, without straying too far in either direction. For a truly random sequence of bits, this path should look like a random walk, fluctuating around the horizontal axis (sea level).

The 'Approximate Entropy Test' measures the randomness of a sequence by comparing the frequency of overlapping blocks of bits. Picture a window, two bits wide, moving along the sequence of ones and zeroes. Count the occurrences of all patterns (00, 01, 10, 11) and draw a histogram with the percentage of occurrences on the *y*-axis. Then widen the window to three bits, slide along the sequence, count the occurrences of the patterns (000, 001, 010, ..., 111) and draw the histogram. Redo the procedure for successively wider windows. If the bars of the histograms for each window size are approximately uniform (the four bars are all about 25% for the two-bit windows, the eight bars of the three-bit window are all about 12.5%, etc.), then the sequence is likely random. Significant deviations from uniformity suggest non-randomness or patterns within the sequence.

In general, if a pattern occurs more frequently than it is supposed to, the sequence fails that particular test. And patterns can be very long before they start repeating. A sequence of seemingly random numbers could be a million digits long...and then start repeating itself; it would be like using RANDs *Million Random Digits* over and over again. Such loops are especially vexing because they invalidate the sequence but are very difficult to spot.

However, the existence of a pattern in a sub-sequence does not necessarily mean that the sequence is non-random. If the sub-sequence 123123123 is part of a million digits long sequence, the entire sequence may very well be random. Hence, the authors of the report add a word of caution:

> ...a tester should note that apparent deviations from randomness may be due to either a poorly designed generator or to anomalies that appear in the binary sequence that is tested (i.e., a certain number of failures is expected in random sequences produced by a particular generator). It is up to the tester to determine the correct interpretation of the test results.

The terms 'anomalies' and 'failures' seem to be misnomers in this context. What the authors must have meant is that sub-sequences that look abnormal are actually quite normal; they must appear at some point in sufficiently long sequences of random numbers.

శ్రీ

Before we conclude this chapter, let me remark about the two kinds of random numbers that we have mentioned earlier and to which we will refer throughout this book. In common parlance, when one speaks of random numbers, one usually means decimal digits between zero and nine. Computer scientists prefer to work with random binary digits (bits), zeroes, and ones. But decimal numbers and binary numbers are equivalent and one can easily be converted into the other (see the Appendix).

శ్రీ

In this chapter we described what random numbers should be like, what properties they must have, what tests they must pass. But do random numbers exist at all? Maybe they are just a figment of the imagination. Or, on a more basic level, does randomness occur at all? We will investigate these rather philosophical questions in the next chapter.

APPENDIX: CONVERTING DECIMAL NUMBERS TO BINARY NUMBERS AND VICE VERSA

DECIMAL TO BINARY

1. **Divide** the number by 2.
2. **Record** the remainder (0 or 1).
3. **Update** the number to the quotient of the division.
4. **Repeat** until the quotient is 0.
5. **Read** the remainders in reverse order.

Example: Convert 13 to binary

- $13 \div 2 = 6$ remainder 1
- $6 \div 2 = 3$ remainder 0
- $3 \div 2 = 1$ remainder 1
- $1 \div 2 = 0$ remainder 1
- Binary: 1101

BINARY TO DECIMAL

1. **Write** down the binary number.
2. **Multiply** each bit by 2 raised to the power of its position index (starting from 0).
3. **Sum** all the results.

Example: Convert 1101 to decimal

- $(1 \times 2^3) + (1 \times 2^2) + (0 \times 2^1) + (1 \times 2^0) = 8 + 4 + 0 + 1 = 13$

CHAPTER 2

To Be or Not To Be?

Do Random Numbers Exist?

Now that we have defined, described, and tested random numbers, the question arises: do random numbers exist? Can true randomness occur in nature? Albert Einstein's famous dictum "God does not play dice" – uttered about the then new-fangled theory of quantum mechanics and its inherent random nature – expressed his belief that everything is deterministic, that there is no true randomness in nature. Did the great physicist turn full circle to the early religious doctrines which claims that it is God's finger that determines everything?

Einstein was no atheist but he was no believer in a traditional God either. Rather, as he wrote in a telegram to a New York rabbi, "I believe in Spinoza's God who reveals Himself in the regular harmony of all that exists…"[1] And according to Einstein, randomness, as expressed, for example, in Heisenberg's Uncertainty Principle, was not part of that harmony. He was convinced that there was more at play than what physicists know. There must exist a hidden determinism that governs subatomic particles, he believed. The argument is difficult to disprove, precisely because the determinism is hidden; after all, it's generally impossible to disprove a negative.

---

[1] *"Ich glaube an Spinozas Gott der sich in gesetzlicher Harmonie des Seienden offenbart."* New York *Times*, 25. April 1929.

DOI: 10.1201/9781003641520-4

How can one rule out the existence of hidden forces or hidden variables? Just because one failed to find them?

But for once Einstein was wrong; randomness, as it occurs at the quantum level, is real. Niels Bohr (1885–1962), Einstein's Nobel Prize winning debating partner, saw the unpredictability of particle behavior as evidence that at the fundamental level of the universe, determinism is replaced by randomness. Though Einstein did not seem to accept it, the fact that elementary particles behave randomly has by now been firmly established. But how about on a macro scale can randomness manifest itself in the observable world for all to see? Or is there a middle path, in between the finger-of-God way of thinking and the belief in pure randomness?

Even the throw of a coin only seems random; if its initial position and all forces, all angles, all external influences were known, the outcome of the throw could be computed. According to this viewpoint, the supposed randomness occurs only because the initial conditions can never be determined precisely enough.

Another school of thought holds that the concept of randomness is not only due to the limits in measurement precision but simply an expression of the limits to our understanding. Whenever something appears to be random, so that thinking goes, there must exist hidden variables that would explain everything … if they were known to us, nothing would be random.

This school of thought says that random phenomena simply occur due to rules too subtle to be understood by human beings. To wit: the convoluted algorithm that produces pseudo-random numbers (which we will discuss in later chapters) plays the role of the hidden variables. The resulting number sequence looks random but reveals itself as completely deterministic as soon as the algorithm is disclosed.

Unfortunately, a hidden variable argument, just like the finger-of-God argument, cannot be refuted in the macro world.[2] Hence, one is at a loss when trying to prove that something is truly random. Phenomena can be unpredictable, chaotic, incompressible…but random?

A question that we must answer at the outset is whether to judge randomness of a number sequence by the features of the sequence itself or

---

2   For hidden variables in the quantum world, see Chapter 11.

by the characteristics of the process by which it was created. Is randomness intrinsic in the sequence or is it a trait of the generating process? In other words, is it the outcome or the process which should determine randomness?

A very long sequence of numbers can be subjected to various statistical tests, but even if it passes all of them, the best one can say is that *non*-randomness has not been proven. For a short sequence, one cannot even do that because non-random-looking sequences could have been generated by a random process. For example, a sub-sequence of the form 000000000000 must appear in a long enough random sequence at some point. In fact, if such a sub-sequence never appears, the sequence would not be truly random.

In that case, the best one can do is to inspect the process. If the process that generated the sequence is random, then even a non-random-looking sub-sequence like 000000000000 should be considered random.[3] But how do we determine whether a generating process is random? Take a coin toss. If the coin is fair, the sequence of heads and tails is random. But how do we determine whether the coin is fair? The fact alone that the two sides, heads and tails, differ in appearance may introduce a bias. It could be that the embossed head is slightly heavier than the picture embossed on the tails side. In principle, only if the two sides are quite indistinguishable can the coin be said to be fair. But then it would be impossible to determine on which side the coin landed. So, the best one can do with a regular coin is to toss it many times and ascertain that the frequencies of both heads and tails are 50%. And with that we are back to inspecting whether the output is random which we already know cannot be decided.

Whichever way we define the meaning of randomness – as a long-run frequency, by the length of the generating algorithm, as a function of our ignorance – there is one characteristic that is common to all definitions: unpredictability. But while unpredictability is a necessary condition for randomness, it does not suffice to define it. As chaos theory shows, even fully deterministic systems can be unpredictable.

So, does randomness exist and if so, can it be proven? The short answers are we don't know, and no it can't. The best we can do is accept its existence

---

[3] Nevertheless, one would not want to use such a sub-sequence to perform Monte Carlo simulations (see Chapter 13). Richard von Mises coined the term *Kollektiv* for sequences whose sub-sequences exhibit the same frequency as the entire sequence (see Chapter 4).

as an axiom. Like the existence of God, it is a matter of belief. Hence, randomness is more of a concept than a natural phenomenon or a concrete feature.

ℳ

Let us return to random number sequences. When confronted with a sequence of numbers, would we be able to determine whether the numbers are random? Would we recognize a sequence of numbers as random? No, we would not! If we were able to do so, we would have to explain that we reached that conclusion either because the sequence is similar to something with which we are familiar, or because we know the cause for the appearance of these particular numbers. But if there is an explanation or a reason for the numbers, the sequence is not random. In short, if we recognize it, it cannot be random.

Hence, we have a paradox on our hands: if the sequence is random, we cannot prove it; if we could prove anything whatsoever about the sequence, it would not be random. So, back to our deeper question: do random number sequences exist at all?

The reason for the paradox is that a sequence of random numbers is characterized not by what it is but by what it is not: not orderly, not predictable, not looping around. In other words, a random number sequence is defined as a negative. And as is well known, one cannot prove a negative: the statement 'elephants exist' can be proven easily, just by pointing to an elephant in the zoo. But to prove that 'unicorns do not exist,' one would have to make an exhaustive search of the universe without finding any unicorns. Hence, it is impossible to prove the non-existence of unicorns.

Similarly with number sequences. To prove, for example, that an infinitely long sequence of numbers is not just a collection of loops, one would have to examine the entire sequence. Even after unsuccessfully inspecting a very long part of the sequence, the end of one loop and the beginning of the next might be lying just around the corner. To prove that random, that is, non-loopy sequences exist, one would have to make an exhaustive search of all infinitely long sequences and find one that has no loop. Hence, obviously, it is impossible to prove the existence of non-loopiness.

Nor can the non-existence of any order or predictability be gleaned by inspecting a finitely long sub-sequence. A search may have been unsuccessful so far, due to the limits of our abilities; if we inspected just a little bit more of the sequence, we could maybe have recognized a regularity.

Or, to argue the other way around, we may have identified a regularity in a sub-sequence and erroneously concluded that the sequence is non-random. But the perceived regularity in a sub-sequence could just be one of the necessary artefacts of a truly random sequence. To conclusively exclude any regularity, we would have to inspect the entire sequence. And even if we could inspect the entire sequence without finding any regularity, we would not be sure that the sequence is random. Maybe we just have not been able to identify the algorithm that produced this particular pseudo-random sequence.

On a metaphysical level, we would have to compare the sequence under scrutiny to everything we know. If it is similar to anything, then it is not random. Therefore, the sequence must be dissimilar to anything we know. So, by definition, we would not recognize a sequence as random, even if we ran into it.

To summarize: since one cannot prove the absence of order, predictability, loopiness, or of anything concrete, the question 'do random number sequences exist?' cannot be answered. It is like asking whether God exists. For believers (and even for Spinoza), the proof is everywhere: nature, animals, human beings. For atheists, proof of God's non-existence is impossible. That leaves the agnostics. For them, God and random number sequences exist as concepts. But in reality…not so much.

The one issue most scientists – with the notable exception of Albert Einstein – seem to agree upon is that quantum phenomena are truly random. But are they really? Just wait and see…

CHAPTER 3

Parapsychology and Psychokinesis

Can Humans Guess, Create, or Recognize Randomness?

As a student in the late 1960s at the Swiss Federal Institute of Technology in Zürich (*Eidgenössische Technische Hochschule*, ETH), this author listened to numerous lectures that were attended by barely a handful of auditors. But there was one interdisciplinary lecture series that filled the *Auditorium Maximum*, the institution's largest, multi-level lecture hall to capacity, with people sitting on the stairs and standing in the aisles. It was the course "Ergebnisse und Probleme der Parapsychologie" (Findings and Problems of Parapsychology) delivered by the Swiss psychiatrist Carl Alfred Meier, a disciple of C.G. Jung, and his successor as professor of psychology at the ETH. As befits lectures at an institute of technology that did not even possess a department of psychology, the lecture was geared towards the quantitative, namely to statistics.

In particular, Meier described experiments that the American botanist James Banks Rhine performed at Duke University in the 1930s. Rhine, who had turned to psychology after obtaining his doctorate, founded the parapsychology lab at Duke and strove to make parapsychology an accepted science. The customary way to achieve this was to perform experiments and to evaluate the results statistically. The subject matter that Rhine was interested in was extra-sensory perception.

DOI: 10.1201/9781003641520-5

ॐ

Parapsychology may be one of the stranger phenomena in the history of science. The reason we talk about this esoteric subject in a book about random numbers is that research into parapsychology utilized random phenomena, ranging from shuffled cards to random numbers generated by radioactive decay. To measure purported psychic abilities, subjects were told to 'guess' the outcomes of random events or the numbers that were generated randomly.

The experiments that James Rhine performed at Duke University, for example, consisted of one person randomly choosing a card from a deck of 25, 5 of which each displayed one of five symbols (a yellow circle, a red cross, blue waves, a black square, a green star) and of the subject trying to guess the symbol that the experimenter had drawn. Many experiments were run, each consisting of many trials and the results were evaluated statistically for their significance. This is what Meier lectured about at the ETH.

On average, without extra-sensory perception, subjects should guess the randomly drawn cards correctly about once every five times. But Rhine obtained some spectacular results. Some of his subjects achieved up to 40% correct guesses, double what they should have if their guesses had been dictated purely by chance. For Rhine and his followers these were strong indications, if not conclusive proof, of the existence of extra-sensory perception.

But the results could never be replicated and soon became suspect. Critics suspected irregularities, if not outright fraud. The colored symbols on the cards could have been reflected in the eyeglasses of the experimenter, the backs of the cheaply produced cards may have shown a faint outline of the symbol printed on the other side, the edges of reused cards were fringed, the tester's facial expression, voice inflection, and breathing patterns may

FIGURE 3.1   Zener cards (From https://commons.wikimedia.org/wiki/File:Zener _cards_(color).svg).

have given subtle clues. Furthermore, Rhine may have neglected to report experiments with negative outcomes.

Key to the experiments was, of course, the randomness with which cards were drawn. But the testing protocols at Duke were not up to the standards that science would require only a few decades later. Before a card was drawn, the deck was handled by the experimenter and often even by the subject, significance levels were overstated, the method of randomizing the cards was suspect.

Extra-sensory perception made for a cool radio show. From September 1937 to January 1938, every Sunday evening at 5 pm, the Zenith Radio Corporation aired a nationwide broadcast across the United States, inviting the public to test their extra-sensory abilities. Ten students who believed in extra-sensory perception were sequestered in a small Chicago studio and received a sequence of five symbols (e.g., heads or tails, white or black, circle or cross, wave or star) that had been randomly chosen through the spins of a roulette wheel. While the students concentrated on the symbols one by one, they 'broadcast' their thoughts nationwide. The audience was asked to determine on which of the two symbols the group of telepathic senders in the studio was concentrating. Several tens of thousands members of the public responded, making more than a million individual guesses.

As the skeptical reader may have expected, no telepathic transmissions were determined… in spite of the Zenith Corporation's valiant efforts to prove otherwise. But the radio show did have an academic interest. The psychologist Louis Goodfellow from Northwestern University noticed that when the panel of students were given a 'popular' sequence to transmit, the audience did exceptionally well. When an 'unpopular' random sequence was broadcast, the results were abysmal.

And what were the popular sequences? Of the 32 possibilities of arranging 5 symbols, the 8 which seemed most random, like 11010, 11001, 10011, or 10010 (and sequences with 1 and 0 interchanged) were chosen in 48.61% of the times, while the 8 least random seeming sequences, 11111, 11110, 10000, 10111 (and their inverses) were chosen by the audience in only 7.4% of the trials.

ᴂ

More rigorous, though still very suspect, were the experiments carried out in the early 1970s by the German physicist Helmut Schmidt, director of the Institute for Parapsychology in Durham, North Carolina. Schmidt's objective was not only to show that people could guess randomly drawn numbers, but that they could actually influence the randomization process with their minds! They could 'pre-guess' which number would come up next, before it was even drawn.

To verify his astounding, truly outlandish claims, Schmidt constructed an ingenious generator of random numbers. The apparatus consisted of a radioactive source (strontium 90), a Geiger counter, and a four-step electronic switch controlling four lamps. He described it thus:

> The Geiger counter responded to ionizing particles emitted from the strontium 90 at random times. A high frequency pulse generator advanced the switch rapidly through the four positions. When a gate between the Geiger counter and the four-step switch opened, the next particle that reached the Geiger counter stopped the switch in one of its four positions and illuminated the lamp corresponding to that position.

This was a generator of honest-to-goodness true random numbers between one and four.

Schmidt's experiments consisted of generating and storing random numbers, without anybody seeing them. The idea was that though the numbers now existed, they could still turn out to be anything, as long as nobody had perceived them. At some later date, he would ask participants to think intensely about the numbers that were about to be revealed and then look at them. The intention was to show that participants could 'influence' random sequences that had been generated months before, just by vigorously thinking about them.[1]

Schmidt coined the term 'retrocausal-psychokinesis' for the supposed phenomenon which signifies a phenomenon doubly abstruse: not only a psychic ability to cause something to happen just by mental effort, but to cause it to happen *after* it has already occurred. Schmidt even used experts at mentalism as subjects, like adepts of meditation, or masters of the martial arts. Surely, if anybody had psychic powers it must be they. But the evidence for retrocausal-psychokinesis in his experiments bordered on the insignificant. After all, deviations of 1% or 2% above the expected level are bound to occur from time to time when the number of trials is very large.

Consequently, Schmidt's entire body of research remained extremely controversial, with accusations of fraud and trickery not ruled out. What

---

[1] Schmidt may have been inspired by the theory of quantum mechanics which states that a particle's state would only be determined once it was observed.

remains is his random number generator, patented several years later…by others.

୬

But that was not the end of explorations into psychokinesis. In 1979, Robert G. Jahn, the Dean of Engineering at Princeton University, established the Princeton Engineering Anomalies Research (PEAR) group. Building on Schmidt's research, PEAR researchers investigated whether subjects could affect the outcome of events that should, in principle, be random. They created a device on a big wall that dropped 9,000 balls randomly through a matrix of 330 pegs, scattering them into 19 bins at the bottom. By the rules of chance, a majority of balls should land in the middle cylinder, and the numbers should taper off towards both ends according to the so-called Gaussian normal distribution (the bell curve). Subjects were asked to mentally direct the balls to the right or to the left of center.

Another experiment was conducted with a 'random event generator,' a commercial microelectronic noise source, whose output was transcribed by appropriate circuitry into a random sequence of zeros and ones. By the rules of chance, this event generator should produce about 50% each of the two bits. In the experiment, subjects were asked to skew the bits towards zero or one.

Though in the ears of academics, the label 'Princeton' usually invokes adjectives like excellent, rigorous, impeccable, the results achieved by the PEAR were underwhelming. Minute differences were styled up to astronomical significance levels though this said nothing about the results' true significance; with huge number of trials, even minuscule differences automatically acquire artificially high significance levels.

One of PEAR's survey papers is indicative of the dubious claims: "Some operators achieve PK [psychokinesis] results in only one direction, some in neither, some in both, and some show inverted results." Critics asserted that PEAR's research lacked scientific rigor, used poor methodology, and misused statistics. Eventually the lab was considered an embarrassment to Princeton University, and in 2007 it shut down.[2]

Actually, the lack of experimental verification of psychokinesis, regular or retrocausal, is very fortunate. A supposed ability of psychokinesis to manipulate experimental results would undermine the very

---

[2] It fused with the similarly dubious International Consciousness Research Laboratories.

foundation of randomness as we know it. For if it were a real phenomenon – if random events and numbers could be directed and skewed by mental processes – we would have to forget about random number generators altogether. Traditional methods that ensure unpredictability and security for cryptography, statistical sampling, and various scientific and commercial simulations would no longer work.[3]

Since the evidence shows that humans do not have extra-sensory mental powers that would allow them to steer random numbers that were generated by mechanical or electronic gadgets in one direction or another, we now turn to the question whether people can produce from their minds sequences of numbers that are random. Can human beings serve as random number generators?

The short answer is no. Sequences that people typically produce differ in fairly systematic ways from those that are produced by truly random processes.

The long answer is 'it depends.' It depends on how the notion of randomness was defined in the instructions to the subjects, and which tests were then used to evaluate the randomness of the sequences that the subjects produced, for example, frequencies of alternations from one to zero and vice versa; pair, triplet, or n-tuple repetitions; autocorrelations; etc.

In the 1930s, the German philosopher of science Hans Reichenbach already noted that

> Persons not acquainted with mathematics often ... are astonished at the clustering that occurs...If a person not trained in the theory of probability were asked to construct artificially a series of events that seems to him to be well shuffled, there would not be enough runs in it.

As a result, sequences produced by humans would overstate the frequency with which outcomes alternate.

Hence, as can be expected, people tend to avoid runs of numbers that they believe – according to their understanding of randomness – would be considered non-random. When asked to produce random binary sequences,

---

[3] To exclude researchers' influence on the outcomes of clinical trials – be it subconscious or parapsychological – serious scientists always conduct 'double-blind' experiments.

people provide excessively balanced numbers of ones and zeros and too few repetitions. In general, they tend to avoid any kind of regularities, even though in truly random sequence such runs would actually appear.

In a later study, two researchers found that, by and large, people exhibit three characteristics: (1) a tendency to think that regular sub-sequences (e.g., 1111, 0101) occur less often than irregular sub-sequences (e.g., 0100, 1101), (2) an expectation of equal numbers of zeros and ones within a sequence, and (3) a tendency to overalternate between outcomes when generating random sequences.

A hilarious Peanuts cartoon in 1952 put it neatly. Linus is sitting for a true/false exam and muses:

Let's see now. In a true or false test, the first question is almost always true. That means the next one will be false to sort of balance the true one. The next one will also be false to break the pattern. Then another true and then two more false ones and then three trues in a row some place...then another false and another true...

Presumably Linus aced the test with his answer sequence 10010011101. His conclusion: "If you're smart you can pass a true or false test without being smart!"

❧

The inability to produce random number sequences spontaneously begs the question, whether people can be trained to do so. One experiment purported to show that they can. In order to establish a baseline, subjects were first asked to generate a random sequence of bits, without any additional instructions. Next, after receiving feedback as to how well their sequences conformed to specific randomness criteria, they were asked to create new sequences. The feedback sessions lasted for an average of six hours and consisted of hundreds of training runs. Subjects who managed to create sequences that approximated randomness were paid a $15 premium. According to the criteria established by the author, the study was successful: as the number of feedback trials increased, the sequences produced by the subjects did indeed tend (ever so slightly) towards randomness.

❧

In 1964, Justice Potter Stewart of the United States Supreme Court famously remarked that he could not define pornography "but I recognize it when I see it." So we ask, can people recognize a random sequence when they see it?

One can turn the question around and ask whether people recognize patterns that are hidden in noise like, on the one hand, radiologists who try to detect tumors in blurry x-ray images, or, on the other hand, like children who think they discern fairies in cloud formations.

In one experiment, two sequences of 150 zeros and ones were shown to the subjects; one sequence had been designed so that two- and three-item sub-sequences occurred with the frequencies that would be expected in a series of coin tosses (a so-called Bernoulli process), the other one not. Surprisingly, a majority of the subjects judged the second, non-random, sequence as more random.

In another experiment, subjects considered repetitions in bit sequences (1 followed by 1, or 0 followed by 0) as less consistent with a random process than alternations (0 followed by 1 and vice versa), especially if the sequences were short. This is an example of the well-known *Gambler's Fallacy* which expresses many gamblers' convictions that after several tosses of heads there must follow tails, even though the probability of this happening remains at 50%. Hence, alternations (changing from heads to tails and vice versa) are judged as more indicative of randomness than repetitions (repeating a head or tail), even if the alternations are more numerous than they would have been in a random sequence.

In experiments with coin tosses, people usually judge the sequence HTHTHTHTHTHT as unlikely, while sequences like THTTHTHHTTTH are considered random. Though we know that both sequences are equally likely, the alternating sequence of heads and tails does not seem random to most people. The reason, according to the two Israeli psychologists Daniel Kahneman and Amos Tversky is that a sequence of coin tosses that contains an obvious regularity is not representative of randomness.

> People view chance as unpredictable but essentially fair. ... they expect even short sequences of coin tosses to include about the same number of heads and tails. More generally, a representative sample is one in which the essential characteristics of the parent population are represented not only globally in the entire sample, but also locally in each of its parts. A sample that is locally representative, however, deviates systematically from chance expectations: it contains too many alternations and too few clusters.[4]

---

[4] Von Mises stipulated that all sub-sequences of an infinitely long random sequence must possess the same characteristics.

In 2001, this prompted a psychology student and his thesis advisor at Stanford University to subject the Zenith Radio data to a new analysis, again not to prove or disprove extra-sensory perception, but to explain for which sequences people have a preference and why. After all, the participants believed correctly that the sequences transmitted over the air were random; so their answers should indicate what kind of sequences they considered to exhibit randomness.

Following Kahneman and Tversky, the two psychologists surmised that when producing such sequences, people pay attention to how representative of randomness the sequence would be. Therefore, given the previous sequence of bits, the next bit is chosen such that the resulting sequence becomes as representative of randomness as possible. Hence, the sequence 0100 should be followed by a 1 (to produce 01001, which seems quite random), rather than a zero (01000, which looks less random). And indeed, of more than 20,000 participants, about three times as many listeners decided that 0100 would be followed by a one rather than by a zero.

∾

Over the decades, numerous experiments have been run about the ability of humans to generate random numbers. Of the many, I cite as illustrations just two papers that were published in the journal *Medical Hypotheses*. A scientist at the University of Cambridge asked several subjects to produce random numbers between zero and nine. The mean of the numbers they generated was 4.501…, just off the theoretically expected 4.500000. Furthermore, a statistical test (the χ^2 test) showed that the digits were close to uniformly distributed. From this, the researcher concluded that people have the ability to let random numbers spring from their minds; he published a paper with the title "Humans Can Consciously Generate Random Number Sequences."

Several years later, three Polish scientists, ran the same experiment but subjected the results to a more rigorous analysis. In particular, they investigated how often a digit was followed by the same digit. Instead of the expected 10% of pairs, the subjects' numbers exhibited only 7.5% of pairs. The researchers also checked how often a digit N, was followed by the digit $N + 1$ or by the digit $N - 1$. Had the number sequences been random, this should have happened in 9% of the cases, but in their experiment, it occurred in about 16%. They expressed their conclusion succinctly in the title of their paper: "Humans Cannot Consciously Generate Random Number Sequences."

By now, the conclusion that human beings are unable to generate random numbers is well established. So, some scientists took their research in a different direction: they exploited the very inability of humans to generate random numbers for new purposes.

A team of seven scientists at the University of Hong Kong, for example, noted that the random number generation performance of a group of schizophrenic people was significantly worse than that of the control group. They linked the deficits in random number generation by people suffering from early schizophrenia to their tendency to exhibit more stereotyped response sequences.

Finally, two Iranian biomedical researchers suggested that the specific characteristics of non-randomness of a human-generated number sequence may be used as a biometric feature to identify the individual who produced it.

❦

In general, people are bad at perceiving or excluding randomness in data, because they tend to find meaning where there is none and erroneously believe that they detected structure instead of randomness. Hence, they often judge a series as non-random because they imagine seeing a pattern that does not truly exist. (And even if there is a pattern, let us recall that patterns do appear in random series.) As Kahneman and Tversky famously observed, people's reasoning about randomness is not guided by laws of probability but by rules of thumb.

A conclusion, formulated more than half a century ago, says it best: "People have a very poor conception of randomness; they don't recognize it when they see it and they cannot produce it when they try."

CHAPTER 4

Nihil Sine Ratione

Entropy, Complexity, Compressibility

It took several millennia for humanity to realize that events may occur randomly, without reason, without cause. Although people had been throwing dice for gaming and divination since the Bronze Age, they mostly believed that the cubes' paths were directed by an omnipotent being. Eventually, the ancient Greeks and the Romans came to realize that the world was partly determined by chance. Though they believed that gods and goddesses had influence over the course of events – for example, by intervening with the throwing of dice – they were only higher beings with superhuman powers, not omnipotent entities who controlled everything.

One of the earliest thinkers who recognized randomness as what it is was Aristotle (384–322 BCE). Classifying natural events into three categories – those that occur necessarily, those which occur usually, and those that occur fortuitously – he characterized the first two as having antecedent causes. The last do not. Hence, he wrote, fortuitous events are not amenable to science since phenomena can be explained scientifically only if they occur necessarily or usually. Accidental occurrences like, for example, a frosty day in the middle of summer, are neither. Therefore "a science of the accidental is not possible," he proclaimed. More specifically, "the causes

DOI: 10.1201/9781003641520-6

from which lucky [random] results might happen are indeterminate; hence luck [randomness] is inscrutable to human calculation."

Aristotle was the first thinker to take randomness out of the hands of the gods and into the realms of mathematical reasoning. Furthermore, his insight forged a link from the abstract concept of randomness to its concrete numerical expression – namely to results that are 'inscrutable to human calculation,' or, in other words, to random numbers. Our quest to grasp the notion of randomness is a continuation of the struggle with the 'indeterminate causes' that Aristotle deemed beyond precise human understanding. But we can do better than Aristotle thought: to comprehend randomness writ large we must simply look through the lens of random numbers.

&

With the advent of Christianity, Aristotle's worldview about randomness went into hibernation. To church fathers and devout Christians, nothing was random, nothing happened by chance, the finger of God was ubiquitous. Every event had to have a cause. If none could be found, it was only because the cause was hidden. Apparent randomness signified nothing other than man's ignorance of the true cause. In the words of the fourth century CE theologian and philosopher St. Augustine, "those causes that are said to be by chance are not non-existent but hidden; we attribute them to the true God." So, not only did randomness not exist, but anybody who pretended that it did was branded a heretic.

No wonder then, that for more than a thousand years, hardly any progress was made in the understanding of phenomena that were described as accidents, lucky breaks, *hasards*, unlucky events, coincidences. Johannes Kepler (1571–1630), the astronomer who discovered planetary motion and was one of the most important representatives of the Scientific Revolution, believed that the universe was deterministic, governed by mathematical laws, leaving little room for true randomness. He had this to say about randomness: "But what is coincidence? Nothing but an idol and the most detestable of idols; nothing but contempt for supreme and omnipotent God, as well as for the perfect world, issued by his hands."

The French mathematician Pierre-Simon Laplace (1749–1827), on the other hand, author of the five-volume *Traité de mécanique céleste* – next to Newton's *Principia*, the most important early work about natural science – did not hold God responsible for phenomena that seem random. He also believed that all events must have a cause, but God was not it. The term 'hasard', he wrote, "only expresses our ignorance about the way in which various parts of a phenomenon organize themselves with each other and with Nature." When Emperor Napoleon enquired where God figured in his works, the mathematician replied dismissively, *Sire, je n'ai pas eu besoin de cette hypothèse-là* ('Sir, I had no need for this here hypothesis.')[1]

And in *The Origin of Species*, Charles Darwin (1809–1882) opined that "I have hitherto sometimes spoken as if the variations…had been due to chance. This, of course, is a wholly incorrect expression, but it serves to acknowledge plainly our ignorance of the cause of each particular variation."

In the short story *The Lottery of Babylon* by Jorge Luis Borges, published in 1941, at the height of Nazism, Fascism, and Stalinism, a mysterious 'Company' issues lottery tickets that can be bought for a copper coin, in the hope of winning a silver coin. After a while, the tickets carry not only prizes but also fines that can be converted into prison terms. Eventually all citizens become participants in an all-encompassing lottery that covers all aspects of life and death. An allegory to God or to the totalitarian state, the question arises how much control people really have over their lives. Does random chance control events or is someone or something pulling the strings?

Gradual recognition that we live in a world where some events occur without a cause eventually gave rise to the notion of randomness. It trickled through concurrently with the theory of probability, as developed by the French mathematicians Pierre Fermat (1607–1665) and Blaise Pascal (1623–1662) in answer to questions about gambling.

The German philosopher and mathematician Gottfied Wilhelm Leibniz (1646–1716), influenced by the work of his French contemporaries, took issue with randomness. As an enlightenment thinker, he was a firm believer in the 'Principle of Sufficient Reason,' succinctly defined by the Latin statement, formulated already by pre-Socratic thinkers, *nihil sine ratione* ('nothing happens without reason'). It accorded with the Newtonian worldview according to which the future state and trajectory of an object can be predicted if its initial state and all forces that act on it are known. Thus, for example, the face on which a die will fall could be predicted. Of

---

[1] By the way, Laplace's wisecrack was not an argument for atheism. It was not God's existence that Laplace had treated as a hypothesis, but merely that God's intervention was not required to explain celestial mechanics.

course, this would require that the initial state and the forces be known to an infinitesimal accuracy since even the smallest imprecision would overthrow all calculations. This is, after all, the famous butterfly effect of chaos theory.

Infinitesimal accuracy is no problem for God, of course, and according to Leibniz, God does recognize the truth of "coincidental things whose complete proof transcends every finite understanding," even if matters or things seem random. Nevertheless, Leibniz conceded in *Discours de métaphysique* that even though everything conforms to the universal order, if the explanation of an event is extremely complex, mortals may regard the event itself as random.[2]

In a seemingly unrelated academic field, the physicist Rudolf Clausius (1822–1888) studied the thermodynamic dissipation of energy, for example, the transformation of energy into useless heat due to friction. He introduced the term entropy for the quantity of energy no longer available to produce physical work. This somewhat abstract concept would soon turn out to be of immense importance for probability theory, information theory, dynamical systems, quantum mechanics …and random numbers. Clausius required the concept of entropy in order to formulate what would become known as the Second Law of Thermodynamics; it says that when energy in a closed system changes from one form to another, entropy increases.[3]

Since probability theory had by now become mainstream, the awareness of randomness, that is, disorder, as a natural phenomenon gained traction. And this is where entropy enters. A decade after formulation of the Second Law, the Austrian physicist Ludwig Boltzmann (1844–1906) offered a microscopic explanation (trajectories and collisions of atoms and molecules) of Clausius's macroscopic observations (energy and heat of a body). Investigating the dynamics of a gas, Boltzmann provided a statistical interpretation of the behavior of atoms and molecules… though their very existence was still only a hypothesis at that time. Namely, when multitudes of particles shoot around in all directions and collide among themselves and push each other out of the way, the state of the gas becomes more and more disordered.[4]

---

[2] With this, he already expressed the basic idea of how computers would produce pseudo-random numbers four centuries later.

[3] The First Law of Thermodynamics says that energy is conserved.

[4] The introduction of entropy to science was truly an international endeavor: apart from Clausius, the German, and Boltzmann, the Austrian, there were also the Scottish scientist James Clerk Maxwell and the American scientist Josiah Willard Gibbs who gave entropy a statistical basis.

And here's the *clou*: entropy, the notion introduced by Clausius to describe unusable energy, is also a measure of the disorder in the gas. Boltzmann developed an equation that defines entropy as proportional to the logarithm of the number of microstates that such a gas could occupy.[5]

A few things about entropy: the higher the disorder, the higher the entropy; low entropy means an ordered state. In other words, entropy is a measure of randomness. If a system is left alone, its entropy can only increase. (That's the Second Law.) The most likely state of a system is the one with highest entropy, that is, the most disordered state.

And the conclusion: since entropy is a measure of disorder, it offers itself as a measure of how random a random number sequence is.

∽

The first modern mathematician to recognize randomness for what it is, was Émil Borel (1871–1956). Though his pronouncement "Whatever the progress of human knowledge, there will always be room for ignorance, hence for randomness and probability" harkens back to the Middle Ages, though 'ignorance' here refers to unpredictability, rather than to hidden causes knowable only by God.

Richard von Mises (1883–1953), an Austrian scientist who fled the Nazis and become professor of aerodynamics and applied mathematics at Harvard, coined the term *Kollektiv* (German for 'collective') for sequences of numbers that are sufficiently long and that exhibit two characteristics: first, as the sequence grows longer, the frequency of the numbers must converge to a limit, reflecting the underlying probability distribution, and the order in which the symbols appear within the sequence is *regellos* (without rule), that is, random. Second, all sub-sequences must possess the same characteristics. (The limit depends on the probability distribution which might be 50–50 for a fair coin, but could vary for different underlying distributions.)

An illustrative aspect of von Mises' definition of random numbers sequences – expressed more rigorously by the logician Alonzo Church (1903–1995) – is that no gambling systems may exist within it. In other words, a sequence of numbers will be defined as random, if it contains no sub-sequence that could be used to improve one's odds of winning.

---

[5] $S = k \ln W$, where k, the proportionality factor, is the Boltzmann constant (1.38×10^{-23} Joule/degrees Kelvin), and W is the number of microstates.

Analogously to Boltzmann's definition of entropy, the electrical engineer Claude Shannon (1916–2001) developed the concept of *information entropy* in 1948 at Bell Labs. It is a measure of the surprise (i.e., the information content) that a message conveys. The higher a message's entropy, the more surprising it is and the more information it is likely to contain. Or, vice versa: the more random a sequence of numbers, the higher its information entropy. For his path-breaking work on information theory, Shannon, later at MIT and the Institute for Advanced Study at Princeton, is considered the 'father of information theory.'

అ

In the 1960s, the Soviet mathematician Andrey Nikolaevich Kolmogorov (1903–1987) weighed in. Famous since the 1930s for having put probability theory on an axiomatic, and therefore mathematically rigorous setting, he took issue with the so-called frequentist definition of random sequences that von Mises had proposed. In particular, the requirement that the frequency of the numbers converge to a limit as the sequence grows longer and longer was too vague for his taste. After all, von Mises was an engineer for whom notions like 'very long' had a satisfactory meaning. But strictly speaking, von Mises' definition is valid only for sequences that are infinitely long. Hence, for a theoretician like Kolmogorov, the formulation of concepts without a precise definition of 'sequences which are sufficiently long' was not good enough.

He spelled it out in a paper he published in an Indian journal of statistics:

> The frequency concept based on the notion of limiting frequency as the number of trials increases to infinity, does not contribute anything to substantiate the applicability of the results of probability theory to real practical problems where we have always to deal with a finite number of trials.

At first, Kolmogorov believed that the problem was insurmountable; the frequency concept, as applied to finite sequences of numbers, would not allow a rigorous mathematical exposition of random numbers. But eventually he came to realize that "the concept of random distribution of a property in a large finite population can have a strict formal mathematical exposition."

First of all, on the negative side, unless the frequency is identical for every sub-sequence, the sequence itself does not represent random numbers. No matter how a subset is chosen – for example, every seventh entry, entries

FIGURE 4.1   Andrey Nikolaevich Kolmogorov (From https://commons.wikime
dia.org/wiki/File:Kolmogorov_Andrey.png).

number 2592 to 3067, the set of even-numbered entries such that the odd-numbered entry just preceding them are 'ones', … – if the frequency in just one subset deviates from the others, the entire sequence must be rejected.

On the positive side, Kolmogorov did provide an ingenious definition for randomness of finitely long number sequences. He considered a finitely long sequence as random if the shortest way to describe it is to repeat it. Let us investigate this a bit further. In an infinitely long random sequence, a 60-bit sub-sequence of 30 ones, followed by 30 zeros,

A: 111111111111111111111111111111000000000000000000000000000000,

is bound to occur in an infinitely long random sequence, with a probability of 2^{-60}. On the other hand, the 60-bit sub-sequence

B: 100101110101000111010100011101001011101010010111011000101010,

also appears with probability 2^{-60} in an infinitely long random sequence. Thus, in principle, both sub-sequences should be considered random because they derive from an infinitely long random sequence. But obviously, sub-sequence B 'looks' random, while sub-sequence A does not. How can that be expressed mathematically?

The answer is that sub-sequence A can be described concisely as '30 ones, 30 zeros' which requires only 17 characters and is, therefore, substantially shorter than the sequence itself. Hence, by Kolmogorov's definition, it is not random. On the other hand, there is no short way to describe sub-sequence B, other than to repeat it bit by bit, using a total of 60 characters. Hence, by Kolmogorov's definition, it is random.

The length of the shortest computer program that can produce a certain sequence is called the Kolmogorov complexity, a.k.a. the algorithmic complexity. For example, the very long string '1 2 3 4 5… .10,000,000' has very low complexity since the simple description 'start with one, and keep appending the next integer until you hit ten million' suffices to produce that string.

The concept was developed in the mid-1960s by Kolmogorov, by the American mathematicians Ray Solomonoff (1926–2009) and Gregory Chaitin (born 1947), and by the Swedish mathematician Per Martin-Löf (born 1942), Kolmogorov's PhD student. Instead of defining randomness probabilistically, they defined it computationally.

Hence, if a string has some kind of regularity, then it can be compressed; and a compressible string should not be thought of as a sequence of random numbers. A string is algorithmically incompressible if its shortest description is the string itself.

Since written text, music, video, graphic art, are not random, algorithms like WinZip, JPEG, RAR, MPEG manage to compress files on personal computers and laptops to a fraction of their initial sizes.[6] And this provides us with another definition of a random sequence: if it cannot be compressed, it is random. In other words, if the shortest computer program that produces a sequence is approximately as long as the sequence itself, the sequence is said to be random. A sequence of 0's and 1's becomes increasingly random, the longer the shortest computer program that generates it becomes.

Now comes the confusing part: the more random a sequence of numbers, letters, pixels, musical notes, the less it can be compressed and the higher its entropy. Does that mean that it contains lots of information? Shakespeare's collected works can be compressed much more than gibberish produced by monkeys pounding away on keyboards. *Gone with the Wind* has less entropy than snow on TV sets, Beethoven's Ninth less than radio static! So, do monkey-gibberish, TV-snow, radio static have informational content?

Well, no! Though a high degree of entropy can indicate the presence of information, there is no direct relationship between entropy and what we consider informational content. The amount of information in a signal or message also depends on its structure and patterns. While high entropy suggests some level of informational content, it does not imply the presence of *meaningful* or *useful* information which refers to the subjective value, relevance, or usefulness of that information in a specific context. Plays, films, the symphonies, have low Shannon entropy and can be compressed, yet they have meaningful informational content. Conversely, white noise is completely random, has high entropy, and cannot be compressed; it has no meaningful information at all. Likewise, a sequence of random numbers, lacking structure or patterns, has high entropy but contains no information … except for the sequence itself.

෴

If you thought that this is confusing, then see what comes next. At the turn of the twentieth century, George Godfrey Berry (1867–1928), a part-time librarian at Oxford's Bodleian Library, was puzzling over something

---

[6] Strictly speaking, JPEG and MPEG use *lossy* compression: first, they remove information that humans typically do not notice, and then they compress the remaining data using methods that reflect Kolmogorov complexity. *Lossless* algorithms include general-purpose compressors like WinZip and RAR, and FLAC (audio), PNG (graphics).

that seemed to him to be a paradox. It so mystified him, that he wrote a letter to the distinguished Cambridge philosopher Bertrand Russell. After several more exchanges, Russell published the paradox in 1908 in the *American Journal of Mathematics*. After that, Berry himself would have descended into obscurity again, had Russell not noted in a footnote that the paradox had been suggested to him by the librarian. Thus, Berry is today remembered by the paradox that has since then been carrying his name.

What was it that so perplexed Berry?

The integer 19 can be referred to in six words as 'the largest prime number below 20.' For the integer 42 we could use four words, to say 'the fifth Catalan number.' Or we can express 1001 in eight words as 'the smallest odd integer greater than a thousand.' How about 'the smallest integer that cannot be expressed in less than a hundred words?' Is this a legitimate description of an integer?

Since the number of words in the English language is finite, only a finite number of integers can be expressed in less than a hundred words. But the total number of integers is infinite. So, there remain infinitely many integers that cannot be expressed in less than a hundred words. And among these impossible-to-express-in-less-than-a-hundred-words integers, there must be a smallest one. Per requirement, more than a hundred words are needed to express it. But the phrase "the smallest positive integer that cannot be expressed in less than a hundred words" – which describes exactly this integer – has only 14 words.

This is not an arithmetic contradiction but a paradox that exposes the limits of language and the dangers of self-reference. The Berry paradox shows that not everything that can be expressed can be consistently interpreted. Hence, phrases like "the smallest positive integer that cannot be expressed in less than a hundred words" are paradoxical because they lead to logical inconsistency when interpreted literally. One would be "describing something, using a description whose form is in apparent contradiction with the meaning of the description."

❧

Beyond exposing the limits of language, Berry's paradox also illustrates the limitations of Kolmogorov complexity as a measure of randomness. On the one hand, the elusive integer can be compressed to its 14-word description ('the smallest integer' … etc.). On the other hand, by its very definition, it "…cannot be expressed in less than a hundred words." Hence it cannot be assigned a definite Kolmogorov complexity.

The paradox highlights the inherent contradictions in defining and measuring randomness and complexity, particularly when applied to random numbers. Self-referential definitions, like those in Berry's paradox, can undermine the rigorous application of Kolmogorov complexity to random numbers, suggesting that our understanding of randomness is inherently limited.

Why is that? Well, let us assume for the moment that a method existed that is able to compute Kolmogorov complexity for any arbitrary string. Let us also say that the algorithm that does this is of size Q. Then this algorithm could be utilized to search – one by one – for the first string it encounters that has a complexity that is larger than Q. Let's denote the complexity of this string by R. But this is impossible since, first, by the definition of Kolmogorov complexity, a string of complexity R cannot be reduced to less than R, but, second, it would have been produced by an algorithm which has a complexity of only Q ($<R$).

We conclude that Kolmogorov complexity, the number of bits of the shortest program that can print a string, is a theoretical concept that is uncomputable; no string can be reduced all the way down to its Kolmogorov complexity. Commonly used compression algorithms, like WinZip or JPEG, only approximate Kolmogorov complexity. Berry's paradox illustrates the inherent limitations of language and logic, suggesting that there will always be challenges and potential for further advancements in areas like compression technology.

With respect to random numbers, the uncomputability of Kolmogorov complexity means that we can't always definitively determine whether a string is random or not based on its shortest representation; there are theoretical limits to our ability to measure and define the randomness of sequences of numbers.

≈

As we have seen in Part I of this book, the concept of randomness permeates various disciplines, challenging our understanding of order and predictability. Random numbers serve as the embodiment of randomness. They are not just theoretical constructs; they are practical tools used in various scientific and engineering and even in commercial applications. From cryptographic algorithms that secure our communications to simulations that predict weather patterns or business scenarios, random numbers are indispensable.

Having established an understanding of what random numbers are, and of their significance, we turn our attention in Part II to their myriad applications as powerful tools that influence many aspects of our daily lives. In the chapters that follow, we will delve into ways in which random numbers are utilized, from the gaming tables of casinos to polling and surveys, from the unpredictable movements in financial markets to the random sampling techniques in scientific research. We explore the profound impact of random numbers and uncover how these weird sequences help understand, predict, and control our world.

II

**Random Numbers
What Are They Good For?**

CHAPTER 5

'The Daily Number'

Gaming and Gambling

The player or 'caster' calls a 'main' (that is, any number from five to nine inclusive). He then throws with two dice. If he 'throws in,' or 'nicks,' he wins the sum played for from the banker or 'setter.' Five is a nick to five, six and twelve are nicks to six, seven and eleven to seven, eight and twelve to eight and nine to nine. If the caster 'throws out' by throwing aces, or deuce-ace (called crabs or craps), he loses...

This is how the *Encyclopedia Britannica*, in its 1911 edition, begins to describe *Hazard*, a game of dice very popular at the time in England. Mentioned in Geoffrey Chaucer's *Canterbury Tales* in the fourteenth century, the game continued to be played in the nineteenth century for large stakes at famous gentlemen's clubs in London. The name of this game, which depends not on skill but only on luck and chance, may derive from *az-zahr* which is Arabic for dice. In turn, this would give rise to the word 'hazard' which denotes danger and risk. With somewhat simplified rules, the game became known as *Craps* in the United States.

Most games of luck essentially boil down to random numbers. If coins are thrown, the results are zero or one; when dice are cast, the numbers

DOI: 10.1201/9781003641520-8

range between one and six; when cards are pulled, the values go from ace to king; and when roulette wheels are spun, the numbers fall between 0 and 36.

In the game of *Hazard*, the numbers that appear when two dice are cast vary between 2 (2 ones) and 12 (2 sixes). They appear hap*hazard*ly (pun intended) but not uniformly because they are not equally probable. The number three, for example, can appear in two different ways, one die shows 'one', the other 'two', and vice versa, while the number seven can appear in six different ways: one and six, two and five, three and four, four and three, five and two, six and one. Hence, the number seven is three times as likely to appear as the number three. We shall not delve into the intricacies of *Hazard*; suffice it to say that the caster had only a 49.06% chance of winning.[1]

Hazard was not the first dice game. From times immemorial, soldiers, children, regular citizens, and slaves overcame their boredom with dice, often in the hope of making a buck on the side. In Babylonia, Egypt, India, Ancient Greece, the Roman Empire, dice games were ubiquitous. Since these games were zero-sum – what one player lost, the other players won – the games were fair in a way, though, as *Hazard* shows, the chances of winning were not necessarily fifty-fifty. And since nobody was aware of probability theory, it was God or Lady Fortuna who controlled the outcomes.

Unfortunately, gambling was addictive already then. The Roman historian Tacitus (first century CE), described the practices of ancient Germanic players who, after having lost all their property in dice games, went so far as to stake their very freedom on one last throw of dice and, upon losing again, voluntarily went into slavery. Such all-consuming passion, and the frequent quarrels that arose as a consequence, prompted many powers to be to prohibit gambling, or to permit it only on festive days. In Jewish tradition, for example, gambling was frowned upon except on minor Jewish holidays like Hanukka and Purim. In the early nineteenth century, gambling establishments in England hired 'dice swallowers' whose task it was to gulp down these random number generators, when the police showed up.

Not all dice games that rely on random numbers are meant for monetary gains. Many board-games, like snakes and ladders, ask the players to roll

---

[1] The numbers between 2 and 12 appear randomly, but are not uniformly distributed. On average, in 36 throws, 2 and 11 appear once, 3 and 11 twice, 4 and 10 three times, 5 and 9 four times, 6 and 8 five times, and 7 six times.

dice in order to indicate how many steps they must move along the board. Other games, like backgammon and monopoly, rely on random numbers but combine luck with strategic thinking and skills in decision-making.

And many games rely on randomness, though not on random numbers. Scrabble players pick letters randomly, though these are not uniformly distributed – the bag does not contain as many Xs and Qs as Es and As. And while dominoes depict a number of pips on their faces (including a zero), the pips do not carry a numerical connotation but only indicate a pattern.[2]

❧

There are gambling methods other than throwing dice to produce random numbers. Lotteries, raffles, and sweepstakes usually rely on numbers being pulled from an opaque bag, or on ping-pong balls blown out of a transparent container. One popular device used for lotteries around the world, and observed by millions of spectators on TV, blows air into a bowl filled with, say, 50 ping-pong balls. The bowl has one opening on top, just large enough to let one single ball escape. Each ball carries a number, and after swirling around for a while, one is blown out through the opening. After a few more seconds another ball escapes, and so on. Or there may be several machines which toss and tumble balls simultaneously. The numbers on the balls that are blown out are the winning numbers.

Let's say that at the start there are 50 balls in the container and the lottery requires the player to guess 5 numbers. The chances of guessing the first number is 1 out of 50, the second is 1 out of 49. The probability of guessing all five numbers correctly is 1 to 2,118,760.[3] (There are also prizes for guessing just three or four numbers correctly.) The total payout is usually half of what has been taken in, so players are bound to lose money in the long run. However, national and state-run lotteries organize gambling events in order to raise money for good causes, like educational activities or facilities for senior citizens.

Given the turbulence that the air flow causes in the container, and the chaotic movements of the ping-pong balls as they bump into each other,

---

[2] Let us also mention Contract Bridge, the card game, which is characterized by randomness (which cards one is dealt), by hidden information (the cards held by your partner and by the other team), and by skills (tactics, communication, and memory). Finally, games like checkers, Go, and chess display perfect information (everything is known); there is no random influence.

[3] $50!/5!45!$

the process is truly random; hence the numbers blown out of the bowl are truly random. However, this is correct under one condition: all 50 ping-pong balls must be identical. Except for the numbers printed on them, they must be indistinguishable from each other: each ball must be of a diameter of 40 millimeters and weigh exactly 2.7 grams. And this is not always so.

'The Daily Number' was a popular lottery game in the Commonwealth of Pennsylvania in the 1970s and 1980s. The rules were that three digits had to be guessed correctly. Since the same digit could be guessed more than once, either the expelled ball was returned to the container after the first draw, and again after the second – in statistics, this is called 'sampling with replacement' – or balls were spun in three air-blowing machines simultaneously.

Operated by the Commonwealth of Pennsylvania, the drawings occurred twice a day. On April 24, 1980, about six million TV viewers watched the evening drawing, as three machines spun and twirled ten ping-pong balls each. All three balls that were blown out of the drums that evening showed the number six.

Well, 6–6–6 is a nice-looking number but statistically there is nothing special about it and nobody suspected anything untowards.[4] Such a combination is bound to occur about once in every one-thousand games. But when it turned out that the prize money would amount to a record payout of 3.5 million dollars, the lottery authorities (and bookmakers who stood to lose a lot of money on illegal sidebets) pricked up their ears.[5] And when the 'lucky' winners came to claim their prizes (there were about 7,200 winning tickets at $500 a piece), the officials began to dig further. They quickly noted that a very large number of tickets had been purchased that contained the digits 4 and 6. That was very suspicious indeed. And then there came an anonymous tip from the public: there had been a fraud. Several arrests quickly followed.

The ringleader of what became known as the 'Triple Six Fix' was none other than the Daily Number's TV announcer, a man by the name of Nick Perry. Together with two partners in crime, a lottery official and a stage hand, they had managed to inject all balls except the ones sporting 4s and 6s with latex paint. This ensured that the injected balls would not twirl high enough; only 4s or 6s could escape the drum, thus reducing the one thousand

---

4   Some numerologists believe the number 666 to be the *Number of the Beast* or of the *Antichrist*.
5   A sum of 3.5 million dollars in 1980 corresponds to nearly 12 million dollars in today's prices.

possible combinations of digits to just eight: 444, 446, 464, 466, 664, 646, 644, 666 – not random by any stretch of the imagination. On the fateful evening, before the drawing, the conspirators swapped the official balls with the doctored ones. After the drawing the real balls were swapped back.

Of course, the air-blowing machines produce random digits only if nobody cheats. In the Triple Six Fix, not only were senior citizens cheated, to whom the profits of the lottery were supposed to go, but also the millions of gamblers who had bet on any numbers but 4s and 6s. Nick Perry was sentenced to seven years in prison for fraud.

Sometimes history repeats itself. On December 29, 2010, the winning ticket of the *Hot Lotto* draw, a multi-state lottery, carried a prize of $14,300,000. The tickets of the lottery were 'manual play,' which meant that the purchaser selected his or her own numbers. A random number generator would spit out numbers and if they matched those that the purchaser had written onto the ticket, the prize was his or hers.

The multi-million dollar prize went unclaimed for nearly a year, until a Canadian man came to claim it in the name of an unnamed individual who wanted to remain anonymous. The claim was turned down because the lottery's regulations did not allow anonymous winners. On December 29, 2011, only hours before the ticket was to expire, a representative of a law firm in Des Moines, Iowa, produced the ticket at the lottery headquarters. The attorney acted in the name of a mysterious company incorporated in Belize. The lottery officials again refused to release the prize money.

Given the strong whiff of fraud, the search was on for the ticket's true purchaser. Publication of surveillance footage at a Des Moines convenience store enabled the identification of one Eddie Raymond Tipton. This Eddie Tipton just happened to be employed by the multi-state lottery association. As such, he was actually forbidden to buy lottery tickets. But much worse was subsequently revealed.

Tipton was in fact the lottery's director of information security. His job was to write the programs and install the computers that would pick the numbers. When investigators examined the number generator, they discovered that the algorithm's 'dynamic-link library' (dll) that was used to pick the numbers was not the same as the one that had been verified as legitimate. Instead, the dll had two additional segments of code. The first re-directed the number generator if the drawing occurred on three dates of the year – May 27, November 22, and December 29 – provided these dates were Wednesdays or Saturdays and the drawing was after 8 p.m. If these conditions were met, the second additional code segment would produce numbers that were predictable to anyone familiar with the operations of the number generator, the security system, and the variables of the algorithm.

Guess who was intimately familiar with all these? Eddie Tipton. Using his programming skills and access to the computers that generated the winning numbers, he rigged lotteries in Iowa, Colorado, Wisconsin, Kansas, and Oklahoma, defrauding the organizations by altogether more than $24 million. Simply put: the numbers were nowhere near random.

"I wrote software that included code that allowed me to technically predict winning numbers and I gave those numbers to other individuals who then won the lottery and shared those winnings with me," Tipton said. "… I didn't think that anybody was breaking the law at all by giving numbers away. But I gave the numbers away knowing that someone could win." In 2017, he was handed down a 25-year prison sentence but paroled after 4.5 years.

৯

Roulette is another game of chance, with a ball and a wheel producing random numbers between 0 and 36.[6] To indicate their bets, gamblers place chips onto the cloth-covered table next to the wheel. Bets are possible on a specific number, on reds or blacks, odds or evens, the lower half (1–18), the upper half (19–36), columns, rows, and all kinds of other combinations.

The renowned French mathematician Henri Poincaré (1854–1912) investigated the probability of red or black numbers in 1899. His conclusion was, surprise surprise, that each color must appear half the times. As a postscript he noted that even well-versed players are prone to making an error that is nowadays known as the 'Gambler's Fallacy.' They may believe it worthwhile to place their bets on black after a streak of, say, six consecutive reds, since it is very rare to have a streak of seven consecutive reds. True, wrote Poincaré, it is very rare to see seven reds in a row. But, he added, the probability of six reds followed by a black is just as rare.

By the way, it is reported that the longest streak of one color occurred in an American casino in 1943 when the color red appeared 32 consecutive times. And in 2012, at the Rio Hotel and Casino in Las Vegas, the number 19 occurred 7 times in a row. Actually, this should not be all that surprising: according to the laws of probability, and given the billions and billions of roulette spins all over the world over all these years, such occurrences are bound to happen sometime, somewhere.[7]

---

[6] American roulette wheels also have a double-zero, to increase the casino's odds.

[7] Actually, on average, you would expect to see a specific number appear seven times in a row once in about a hundred billion spins ($37^7 = 94,931,877,133$). Making some assumptions, it should happen somewhere in the world once every 15 years or so.

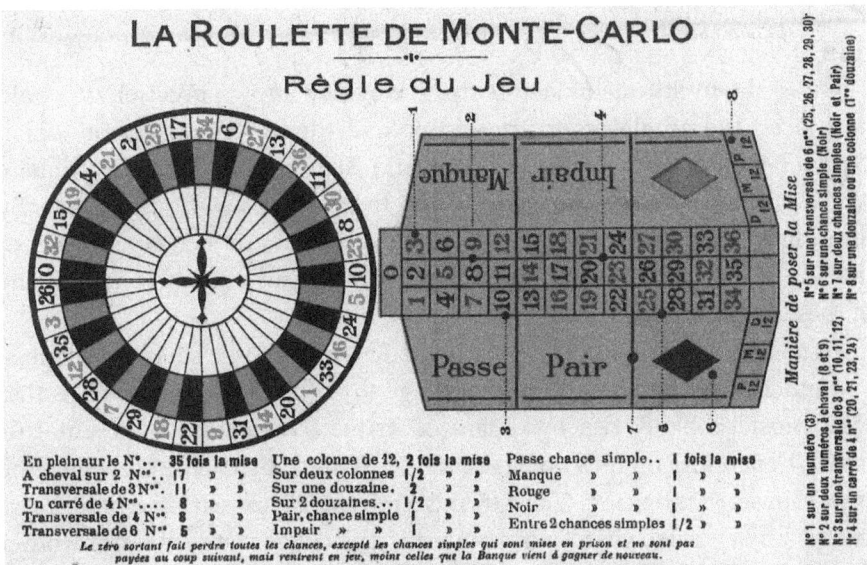

FIGURE 5.1   La Roulette de Monte Carlo 'Rules of the game' (From https://commons.wikimedia.org/wiki/File:La_Roulette_de_Monte-Carlo_R%C3%A8gle_du_Jeu.jpg).

Ultimately, the issue of non-randomness and predictability of the roulette wheel is not a question of mathematics but of physics. If one could accurately measure the speed, direction, and spin of the ball as it is released, the angular velocity of the wheel, the elasticity of the ball and the table, the coefficients of friction, the air resistance, the earth's vibration, and many other parameters which influence the ball's trajectory, one could, in principle, work out in advance into which numbered box the ball will fall. This is exactly what Edward Thorp (born in 1932), a well-known card counter in Blackjack and later professor of mathematics at various universities, did in the 1960s in collaboration with a teammate. This teammate insisted on remaining anonymous; all that Thorp divulged about his associate was that he was a world-renowned scientist.

Key to their system was that casinos allowed chips to be placed on the table after the wheel had been set in motion and while the ball was still rolling, that is, during a time span of about 10 seconds. Using minicomputers, the size of cigarette packs, and micro-switches hidden in their shoes – operated with their great toes – Thorp and Co. timed the ball's first few rotations and let the computers work out its probable trajectory. Signals sent to hidden earphones informed the two players to place their chips just before the croupier announced *"les jeux sont faits"* (in Monte Carlo) or "no more bets" (in Las Vegas).

Given the multitude of factors that have to be known precisely, it would have been too ambitious to try and predict the exact number on which the ball would land. Instead, Thorp and Shannon divided the roulette wheel into eight zones, each comprising four or five neighboring slots. By limiting themselves to trying to predict the octant and betting on each of that octant's four or five numbers, they could expect a payout of $35 if one of these numbers did win.

It took not just knowledge of mathematics and physics, scientific prowess, and much ingenuity, but a lot of courage to try to rip off the casinos that were most probably run by organized crime. Fortunately – or unfortunately, depending on how you look at it – technical problems prevented the two from serious betting… and from serious problems with the mob. Years later it was revealed that Thorpe's mysterious collaborator was none other than MIT's world-famous professor Claude Shannon, the so-called father of information theory. Shannon's interest in predicting the supposedly random meanderings of roulette balls was not surprising. He was, after all, a master of randomness. In fact, he was the creator of the concept of information entropy, the measure of how much randomness a message contains.

And that is where a group of hippies in and around the University of California at Santa Cruz picked up two decades later. The graduate students Doyne Farmer, Norman Packard, James Crutchfield, and Robert Shaw would eventually become known…no, not as successful gamblers but as successful academics. Under the name 'Dynamical Systems Collective,' these then twentysomethings published some of the path-breaking early papers in chaos theory.

But instead of working on their doctoral theses, they seriously contemplated leaving academics to concentrate on getting rich by beating the casinos. Taking time off from school, they spent thousands of hours developing and refining their software and hardware. In trial runs on a roulette wheel that they had purchased, it turned out that while their software was good enough, their hardware ran into similar problems as did Thorp and Shannon's. Nevertheless, as a proof of concept, their shoe computers, micro-switches, and earpieces worked about as well, or as badly, as Thorp and Shannon's. And so, after winning nickels and dimes, they returned to their academic careers.

The upshot of Thorp and Shannon's endeavors, and the accomplishments of the Dynamical Systems Collective, is that – with computing power that was quite moderate compared with today's standards – roulette can be beaten by betting on the octants. But if the winning numbers can be predicted even just to a minor degree, then they are not random! So, actual roulette wheels on actual tables should not serve as sources for random numbers.

∽

To protect casinos from exploitation by computer-savvy gamblers, the State of Nevada codified a law that made it illegal to

> use ... any computerized, electronic, electrical or mechanical device, or any software or hardware, or any combination thereof, which ...

1. Projects the outcome of the game;

2. Keeps track of cards played or cards prepared for play in the game;

3. Analyzes the probability of the occurrence of an event relating to the game; or

4. Analyzes the strategy for playing or betting to be used in the game... .

(Nevada Revised Statutes 465.075)

If that is how the State of Nevada protects casinos from cheating gamblers, how does the state protect gamblers from cheating casinos? In a 'Notice to Licensees,' the Gaming Control Board of the State of Nevada sets out the regulations as they relate to gaming devices that use random numbers:

> Any software random number generator (RNG) used as part of the random selection process must:

1. Not use static seeds upon initialization. The RNG must be seeded using a sufficiently unpredictable value. If a value representing a time or date is used, the value must both be represented in milliseconds or smaller, and must not be viewable anywhere on the gaming device;

2. Continue to cycle when not in use. The RNG must cycle at a minimum of 100 Hz (100 calls per second on average);

3. Only produce outcomes for a single game.

4. Gaming devices such as system-based games or Interactive Gaming Systems that offer multiple simultaneous games or

tables must use a separate RNG and separate initial seed for each RNG.

5. Games that offer RNGs for purposes other than determination of game outcome (e.g. Keno Quick Pick) must use a separate RNG than that used as part of the game random selection process.

According to the 'Notice to Licensees', gaming devices that use a hardware random number generator as part of the random selection process (wheels, ping-pong balls, roulette wheels, etc.) must be able to do the following:

1. Determine if the RNG meets a 95 percent confidence limit using a chi-squared test for goodness of fit…

2. Display a visual notification, clearable by an attendant, if at any time the wheel has failed the chi-squared test for goodness of fit.

3. Disable play in the event that the wheel has failed to meet to the 95 percent confidence limit goodness of fit test for two consecutive testing periods. Attendant interaction is required before enabling the table for play.

4. Provide a report that shows the results of the last chi-squared test as well as the previous 9 chi-squared tests…

Note that in any case it is not in the casinos' interest to rig roulette tables. First, they make their money, and a lot of it, on the ball falling into the zero-slot. Second, if tables are rigged, some players may catch on which would simply cut into the casino's profits. So, why tamper with a winning – and riskless – strategy?

To make sure that both sides really do get it, NRS 465.085 emphasizes that

it is unlawful to mark, alter or otherwise modify any associated equipment or gaming device … in a manner that: … alters the normal criteria of random selection, which affects the operation of a game or which determines the outcome of a game.

So – by the laws of the State of Nevada – gamblers can gamble with confidence, and the casinos – by the law of large numbers – are guaranteed their profits.

❧

When gambling at casinos or betting at lotteries and raffles, all players, except the organizers, are bound to lose in the long run. It does not have to be so. In so-called zero-sum games between two or more opponents, players usually have a fair chance of winning…and of losing. The sum of money that changes hands balances to zero. Early dice-games played by Greek plebeians and Roman soldiers were most certainly zero-sum: whatever one player lost, the other won.

Many games involving dice depend on pure luck; the player who throws the higher number wins. But there are zero-sum games where the outcomes depend on the players' strategies. How to outwit an opponent in a game was the subject of a seminal book written at the Institute for Advanced Studies at Princeton in the late 1940s. The authors were the mathematician John von Neumann, a towering genius of the twentieth century, and the economist Oskar Morgenstern. Their work spawned the Theory of Games, an entirely new and very fruitful field in mathematical economics.

In a two-person zero-sum game, a rational player would try to employ a strategy which minimizes the maximum loss, or maximizes the minimum win. But as soon as the other player realizes what strategy the opponent uses, he or she will answer with a corresponding counter-strategy. And since they are both rational and both know each other's thinking, they usually end up in a so-called Nash-equilibrium:[8] given the opponent's strategy, neither player could have done better by employing a different strategy.

If the game is played several times, however, a player might do better by not following the rational course of action but, rather, by randomizing the strategy. In the rock–paper–scissor game, for example, a naïve player might mindlessly cycle through the three options, or simply choose whatever the opponent chose in the previous round. She or he would be bound to lose every round as soon as the opponent catches on. Therefore, the best strategy is to follow no strategy: choose rock or paper or scissor randomly.

What von Neumann and Morgenstern realized was that unpredictability itself can be employed to gain advantage in certain games: true randomness can confer power over those who, in desperation, seek order when faced with chaos. And what provides that edge? Making decisions based on a sequence of random numbers.

---

[8] Named after the mathematician John Forbes Nash (1928–2015), recipient of the Nobel Prize in economics in 1994, and the Abel Prize in mathematics in 2015.

CHAPTER 6

Choosing by Lot

Sortition, Polling, and Random Sampling

In the Republic of Venice (7th to 18th century), the highest official was the doge:

> Whenever the time came to elect a new doge of Venice, an offi-
> cial went to pray in St. Mark's Basilica, grabbed the first boy he
> could find in the piazza, and took him back to the ducal palace.
> The boy's job was to draw lots to choose an electoral college from
> the members of Venice's grand families, which was the first step in
> a performance that has been called tortuous, ridiculous, and pro-
> found. Here is how it went, more or less unchanged, for five hun-
> dred years, from 1268 until the end of the Venetian Republic.
>
> Thirty electors were chosen by lot, and then a second lottery
> reduced them to nine, who nominated forty candidates in all,
> each of whom had to be approved by at least seven electors in
> order to pass to the next stage. The forty were pruned by lot to
> twelve, who nominated a total of twenty-five, who needed at least
> nine nominations each. The twenty-five were culled to nine, who
> picked an electoral college of forty-five, each with at least seven
> nominations. The forty-five became eleven, who chose a final college

DOI: 10.1201/9781003641520-9

of forty-one. Each member proposed one candidate, all of whom were discussed and, if necessary, examined in person, whereupon each elector cast a vote for every candidate of whom he approved. The candidate with the most approvals was the winner, provided he had been endorsed by at least twenty-five of the forty-one.[1]

The bewildering complexity of this convoluted voting method ensured, for all to see, that corruption and backroom deals play no role in electing the new ruler. In particular, the repeated recourse to the drawing of lots ensured that every person had a chance of being elected (at least, i.e., if he was male and hailed from one of the grand families). The method seems weird, but is the democratic election of a ruler, legislator, mayor, or union president by a majority of voters really superior to the random selection of a leader? Can a crowd of voters identify the best, or at least a suitable, candidate?

Democracy is generally equated with majority rule. Hence, the preferred candidate or the favorite option is the one that is picked by a majority of the electors. 'One man (or better: one person) one vote' and 'the majority decides' are considered sacrosanct rules of democracy. Could choosing office holders by casting lots, or by drawing random numbers, be just as good…or better?

❧

The Venetians were not the first to choose their leader at random. Though the ancient Greeks elected some officials by majority vote, they used lotteries to choose others. The select few who did come to their positions by election were the ones whose jobs required special skills like warfare or money management. After all, generals needed experience and expert knowledge, in order to conduct campaigns and win wars, treasurers had to be wealthy in addition to being savvy, so that public money that they lost, due to mismanagement or through corruption, could be recovered from their personal property. These civil servants, and some others, were elected. Many others were chosen by lot.

… . Neither were the Venetians the last to choose personalities at random. For systems in which politicians and civil servants are selected by lots, the word *lottocracy* has been coined. But public officials aren't the only candidates for random selection. In many countries, juries in criminal and civil trials are chosen at random from among their citizens. During the Vietnam War, recruits for military service were drawn by lot from among the 18-year-olds (see the introduction) random pedestrians may be frisked by police; drivers are tested at random for intoxication, competitors

---

[1] Gottlieb, Anthony, "Win or Lose," *The New Yorker*, July 26, 2010.

at sporting events are drawn randomly for drug testing, green cards are granted to potential immigrants by lottery.[2]

While believers in the 'wisdom of the crowds' implicitly trust the majority to identify and elect the best candidate, there exist socio-political decision-making mechanisms other than majority voting, such as markets, hierarchies, traditions, negotiations and … random selection. Cynics argue that numerous historical examples have proven that crowds can make wrong, even disastrous, decisions. Is the Venetian model of choosing candidates by lottery for political or social service randomly really inferior to the others?

Jean-Jacques Rousseau (1712–1778) pronounced in *The Social Contract* that "Election by lot would have few disadvantages in a real democracy, in which, as equality would everywhere exist in morals and talents as well as in principles and fortunes, it would become almost a matter of indifference who was chosen." A matter of indifference indeed!

The system of choosing office-holders by lottery is commonly known as 'sortition' (from Latin *sortire* 'cast or draw lots'). Every citizen is assigned a number, for example, his or her social security number. When the time comes to fill positions, numbers are drawn at random from a bag or a drum and matched with the jobs that need to be filled. It is supremely important, of course, that everybody has the same probability of being selected.

To choose office-holders by lot carries both advantages and disadvantages; pros and cons must be carefully balanced against each other. While executive positions should be filled with the most qualified candidate and hardly anyone would suggest appointing the American president by a lottery, legislators who represent the underlying population could be chosen randomly.

So, while the executive had better be elected by a democratic majority vote, and law-givers could be chosen by lottery, how should justice be dispensed? Since ancient times, and in many countries today, justice was and is dispensed using both randomness and voting. An American individual, for example, accused of a serious crime, has the constitutional "right to a speedy and public trial, by an impartial jury" (The Sixth Amendment). The thinking is that a collection of regular folks, randomly selected, is best suited to determine whether the accused is guilty or not.

Less accepted, if not downright rejected, is the determination of the actual verdict by a random coin toss. In fact, when a judge in Kentucky

---

2   To prevent terror acts on aircraft, checking only random samples at boarding gates won't do. Terror acts must be prevented altogether, and random sampling would identify the carrier of a bomb only in rare cases. Hence, at airports, everyone, 100% of the travelers, must be checked.

found out that a jury had tossed a coin to decide whether to convict an accused of murder or of manslaughter, he declared a mistrial. And when a judge in Virginia decided on visitation rights in a child custody case by coin toss, he was removed from the bench.

On the other hand, coin tosses and lotteries are often seen as fair procedures in dispute resolutions or in the allocation of scarce goods. To cite just one example, the Bible tells that Moses, commanded by God, instructed the Israelites to "assign this land by lot as an inheritance" (Numbers 34:13). In modern times, lotteries are accepted procedure to assign preferable housing, donate organs for transplantation, allocate spots in fashionable schools, allocate drilling leases, distribute tickets to rock concerts. Lotteries are also used to assign undesirable chores like cleaning services or guard duty and, of course, military draft and jury duty.[3]

If one can envisage selecting legislators randomly, dispensing justice by coin tosses, and allocating real estate by lottery, it seems not farfetched to make business and other decisions by drawing lots. While the Ancients did so because they believed that the gods directed the hand which chose the lots, managers may resort to random drawings for lack of a better system. (For decision-makers too lazy not only to make decisions but even to just toss coins, there are apps on the App store that assist in making random, unbiased choices.)

Julius Caesar's *alea iacta est* (the die is cast), uttered as he crossed the river Rubicon, may be the most famous decision allegedly based on a random toss of a die. (Actually, Caesar did not cast a die to decide whether or not to cross the river; rather his exclamation indicated that there was no way back: once he crossed the river, the choice had been made.)

When business managers are faced with a choice, they will weigh the pros and cons of all alternatives. Problems arise, for example, if several candidates for promotion or applicants for employment are equally qualified. Nepotism would be one, albeit reprehensible, way of choosing; affirmative action, would be a more well-meaning but equally unjust method. A lottery may be the only fair solution. As the saying goes, "sometimes the most rational choice is a random stab in the dark."

---

[3] Though lotteries are considered fair, economists criticize the procedure because it does not allocate goods to those to whom it would give most utility. Auctions or the market place would be more efficient procedures.

While sortition may have its merits for office holders, random sampling is indispensable for data collection. It is one of the basic tools of medical research, scientific investigation, political polling, psychological experiments, quality control, market research. The reason that pollsters make do with samples is that it would be too costly, indeed impossible, to poll all citizens of a nation individually, except for infrequent general elections. Quality controllers examine only a selection of mass-produced widgets because the items to be tested may have to be taken apart and destroyed. Hence only a fraction of the population or of the items is subjected to testing, polling, investigation.

The first question an investigator must answer is what fraction of the population should be scrutinized. How large should the sample be? Statistical theory specifies the parameters that must be taken into account: the size of the universe that is to be investigated, an estimate of the result that one expects, the margins of error one is willing to accept, and the confidence one wants to have that the true result actually lies within the error margins. As we shall now see, it turns out that only the latter two parameters are of importance. (The formula for the minimum sample size and a short discussion is given in Appendix to this chapter.)

Let us say that we want to determine how many of the 10,000 widgets produced daily by a machine are bent upwards, how many downwards. We suspect that it is about half–half and are willing to accept an error of plus or minus 5%. But we want to be 95% confident that the true result lies within that error band. The theory tells us that we should use a sample size of 370.

Now let's estimate the proportions of conservative voters and progressive voters in a nation of 100,000,000. We suspect that about half vote left and half vote right. We are again willing to accept an error of plus or minus 5% and want to be 95% confident that the true result lies within that error band. How large is the sample size? 385.

This is surprising indeed; the size of the required sample is very insensitive to the size of the population. It turns out that if the universe is sufficiently large, its actual size is of negligible importance for the sample size and may be ignored. Only 15 additional entities (widgets or citizens) must be tested when the universe is increased from 10,000 to 100,000,000 (see Appendix).

∽

The second step in sampling is to draw 370 items, or 385 citizens, or whatever sample one needs, from the entire population. To obtain a statistically correct estimate, the procedure must be truly random. And this is where random numbers enter the picture. The fundamental objective of sampling is that each item in the universe that is under investigation must have the same chance of being drawn.

To ascertain, for example, whether a vaccine has a curative effect against a disease, several mice are divided into two groups; one group is administered the drug, the other is given a placebo. Then all mice are exposed to a virus and if a statistically significant number of the vaccinated ones remain healthy while members of the placebo group fall ill, one may surmise that the vaccine prevents infection.

The point that interests us here is how the mice are assigned to the groups. If the investigators of pharmaceutical corporations were allowed to assign the mice according to their whims, they could intentionally or unintentionally assign the healthy-looking ones to the vaccination group and the frail ones to the placebo-group. Then, when members of the placebo group, which were frail from the start, fall ill, while the vaccinated mice don't, Big Pharma could – erroneously or fraudulently – claim that they 'proved' the drug's effectiveness. In order for this not to happen, the assignment of each mouse to one of the two groups must be done completely randomly.

Let us say that there are 100 mice; 50 will be administered the drug, 50 will receive a placebo. The investigators assign numbers from 1 to 100 to the mice and then draw 50 random numbers between 1 and 100. The mice which bear the drawn numbers will be assigned to the placebo group and now the experiment can proceed in an orderly fashion. If the number of vaccinated mice that recovered is significantly higher than the number that recovered from the placebo group, the drug may be deemed to be effective.

There's one more caveat. Even though the mice have been assigned to the two groups in a truly random fashion, the investigators who are hoping for the drug's effectiveness could – wittingly or unwittingly – invest more effort into the well-being of the mice who were inoculated…more stroking, better food, more attention. To prevent this, the investigators themselves must not know which mice received the drug and which got the placebo while the test is in progress. It must therefore be an independent agent who draws the random numbers and assigns the mice to the two groups. Only after all results are in, is it revealed to the investigators which mouse belonged to which group, and the statistical evaluation of the results can proceed.

Such studies are called double-blind. Why double-blind? The researcher is blind to which mice got the drug, but why double? Well, it may be beside the point but the mice also don't know whether they received the real thing or a placebo. However, in pharmacological studies with human patients, both the investigators and the test persons must remain ignorant as to what they were administered. (After all, the very knowledge that a drug has been administered may have a beneficial effect (the notorious placebo effect) on a patient even if the drug itself is ineffective.)

∾

Famous examples of political polling gone wrong because the samples were not random occurred before the American presidential elections in 1936 and 1948. In the first, the magazine *Literary Digest* falsely predicted the Republican contender Alf Landon as the winner over the incumbent Franklin Roosevelt. The magazine had surveyed over two million voters by drawing a huge sample from telephone books, automobile registration lists, and its own readers (it had a circulation of about one million). But it failed to account for one important factor: in the 1930s the owners of cars and telephones, and the subscribers to the *Literary Digest* were overwhelmingly wealthy, well educated…and Republican. In fact, *Literary Digest's* marketing people had compiled the automobile and telephone lists with the express purpose of soliciting further subscriptions to this rather highbrow magazine. Hence, not every American citizen had the same probability of being surveyed; the resulting sample was anything but random.[4] The debacle and subsequent loss of credibility was the main reason for the magazine's demise two years later. Interestingly, George Gallup, the founder of the eponymous polling company, had predicted both the correct winner and the erroneous result of the digest.

Twelve years later, the exact same thing happened again and this time it was Gallup's fault. Based on telephone surveys two weeks earlier, Gallup predicted a win by the Republican Thomas E. Dewey. But it was the Democrat Harry Truman who won. Once again, the owners of telephones who had been sampled were not representative of the electorate.[5] They were wealthier than most American citizens; hence dozens and hundreds of

---

[4] Another explanation for the result was the low response rate. Of the 10 million straw ballots that were mailed out, only 2.3 million were returned, presumably by people who felt so strongly for their preferred candidate that they made the effort to return the ballot.

[5] Another reason for the debacle may have been that when Gallup performed his phone survey, two weeks before the election, 14% of voters were apparently still undecided.

FIGURE 6.1 Dewey defeats Truman (From https://commons.wikimedia.org/wiki/File:Dewey_Defeats_Truman.jpg).

millions of potential voters had no chance of being included in the sample. The sample was not random.

Let us advance to more recent times. Before the US presidential elections in 2016, most opinion polls predicted a victory for Hillary Clinton. Her chances of being elected president were estimated at about 90%. The polls were wrong: Donald Trump won. How could polls, once again, be so wrong? Many reasons were given to explain the polling debacle, I will only cite one: voters with higher education levels were more likely to support Clinton and since better educated people are significantly more likely to participate in surveys than those with less education, Clinton supporters were overrepresented in the samples.

This problem can in general be avoided by correctly weighting the raw polling results. Even when using a random number generator to select responders, pollsters must account for the under-representation of less-educated adults in their weighting. Unfortunately, in the 2016 elections many polls did not adjust their weights to correct for the overrepresentation of college graduates in their surveys.

❧

What if the persons to be polled or the widgets to be tested are picked with the help of a random number generator but the random numbers did not pass all the randomness tests that were described in Chapter 1? For example, what if they were biased towards the odds, as in the first version of the RAND Corporation's million random numbers? Well, it might pose a problem.

Let's say that a conveyor belt produces widgets but that every second one has a defect. If the odd numbers are more frequent in the series of supposedly random numbers, and the widgets to be tested are chosen according to these numbers, the non-defective widgets would be subjected more often to the test than the defective ones, or vice versa. The erroneous result would be that the conveyor belt receives a clean bill of health even though it should not, or vice versa. Only if each element in the universe, each mouse, patient, and widget, has the same probability of being selected, can the results of the random sample be accepted with confidence.

The takeaway for pollsters and testers: random numbers are crucial in order to choose samples in such a way that every citizen, every widget has the same chance of being chosen. But even truly random numbers do not guarantee correct forecasts or results; additional safeguards against unintended biases must be instituted.

APPENDIX

SAMPLE SIZE

Let us denote the size of the entire population by N, an estimate of the result that one expects with p (say 50%), the margins of error one is willing to accept with e (e.g., ±5%), and indicate the confidence one wants to have that the true result actually lies within the error margins by y (e.g., 95%).

Then the sample size is computed as (see any statistics textbook or Google 'sample size'):

$$\text{Sample size} = \frac{\dfrac{z^2 \times p(1-p)}{e^2}}{1 + \left(\dfrac{z^2 \times p(1-p)}{e^2 N} \right)},$$

where z is the value of the so-called normal distribution (the bell curve) at the level y. For the 95% level, $z = 1.96$; for the 99% level, $z = 2.56$. We see that for a large population (e.g., $N > 10,000$), the population size may be ignored. The most conservative sample size (i.e., the largest) results when $p = 0.5$; hence we may as well use this p-value, and the approximated equation for the sample size becomes

$$\text{Sample Size} = \tfrac{1}{4}\, z^2/e^2$$

With error margins ±5%, this gives a sample size of 370 for the 95% confidence level, and 666 for the 99% confidence level.

CHAPTER 7

Making Money

Cognitive Dissonance and Random Walks

Appreciating the power and elusive nature of random numbers requires recognizing that human beings are inherently poor at identifying events governed by randomness. Our brains are wired to find patterns, a trait honed by evolution: selection processes favor those who possess predictive power. Yet this often leads us astray when faced with true randomness: genuine signals, buried within vast amounts of data, are often mistaken for random noise. In some ways, true randomness defies the rules of common sense. This chapter delves into the intriguing relationship between randomness and financial decision-making, illustrating how randomness can influence gains and losses.

Imagine the following game: Peter tosses a coin. If it lands heads, you get two dollars. If it lands tails, Peter tosses again; if it lands heads on the second toss, you get four dollars. If it lands tails again, and falls heads only on the third throw, you get eight dollars. And so on.... If it lands heads for the first time only on the nth toss you get 2^n dollars. Should you be willing to pay ten dollars to participate in the lottery? A hundred? A thousand?

In the middle of the seventeenth century, the philosopher Blaise Pascal in Paris, and the judge Pierre de Fermat in Toulouse, in southern France, concluded that the expected value of an uncertain event is computed by

DOI: 10.1201/9781003641520-10

multiplying the potential values with the probabilities of their occurrence. If a gamble carries a prize of $80 and has 10% odds of winning, the expected payout is $8 (= 0.10 × $80). If there are several prize possibilities, the expected payout is the sum of all possible outcomes, multiplied by their respective probabilities. So, if, in addition to the 10% chance of winning $80, there is also a second prize – a 20% chance of a $30 payout – the expected winnings would be $14 (= 0.10 × $80 + 0.20 × $30). A rational gambler should be willing to pay any amount up to $14 to participate in the gamble.

Hence, the expected win of the coin tossing game must be calculated as follows: the chances of the coin landing heads on the first throw, and your receiving two dollars, is ½. The probability of the coin landing heads only on the second throw, and your receiving four dollars, is ¼, the probability that heads will appear only on the third throw and that you will receive eight dollars, is $1/_8$, and so on. If, say, the coin lands on tails nine times in a row, and heads appears on the tenth throw (TTTTTTTTTH), the payout would be $1,024. The probability of such a series of throws is $1/_{1024}$.

Since the expected payout is the sum of all possible payouts (2, 4, 8, …, 1,024, …) multiplied by the probabilities ($1/_2$, $1/_4$, $1/_8$, …, $1/_{1,024}$, …), the expected win is

$$(2 \times 1/_2) + (4 \times 1/_4) + (8 \times 1/_8) + \dots + (1{,}024 \times 1/_{1{,}024}) + \dots$$

$$= 1 + 1 + 1 + \dots + 1 + \dots$$

Wow! Since a sequence of random bits can have a long sub-sequence of zeros (or ones) following each other, there is a real, if minute, chance that many, many tails are thrown before the first head appears and the series actually never ends. To compute the expected value, infinitely many 1s must be summed and the expected win amounts, shockingly, to infinity! Accordingly, a rational gambler should be willing to pay any amount to participate in this lottery.

Obviously, no rational individual would. The solution to the paradox was proposed by Nikolaus Bernoulli (1687–1795), nephew of Jakob (1655–1705), the author of the founding work on probability, *Ars Conjectandi*. Nikolaus stated that the relevant variable is not the actual payout, but the utility that this payout affords the recipient. After all, the utility of an add-itional dollar is much greater to a pauper than to a millionaire who would hardly notice a one-dollar addition to her wealth. Nikolaus proposed the natural logarithm (denoted by *ln*) of the payout as a measure of its utility. Hence, a payout of $100 would afford a utility of 4.60 'utiles' to a penniless pauper but add only 0.0001 *utiles* to a millionaire's utility. Thus, the expected utility for a pauper of entering the game is

$$(\ln(2)\times1/2)+(\ln(4)\times1/4)+(\ln(8)\times1/8)+...$$
$$+(\ln(1,024)\times1/1,024)+...=2.414...\text{utiles}$$

and he or she should be willing to pay about $11.17 to enter the game (ln(11.17) = 2.414...).

The St. Petersburg coin-tossing game depends on events that occur randomly. Hence, we could analyze it – and games like it – using probability theory. In chess, on the other hand, an opposing player performs moves, and one must strive to outsmart him or her. The opponent's and your own decisions influence the outcome, but the game is deterministic in the sense that nothing depends on chance. Finally, in games like monopoly, backgammon, poker, bridge, both chance events and opposing players' decisions influence the outcome.

ᡇ

The St. Petersburg paradox is an example of how people can be *Fooled by Randomness*. This is the title of a book by the one-time bond trader and celebrated author Nassim Taleb. The book's main thesis is that human beings (in particular of the bond trader variety) do not know how to deal with randomness. Previously, I discussed whether humans are able to produce or recognize sequences of random numbers. Taleb's proposition is more basic: the successful ones among his trader colleagues consistently overestimated their own ability and underestimated the role that good luck played in their purported success, while unsuccessful colleagues tend to attribute their mistakes and losses to fortuitous bad luck; both do not give randomness their proper due.

One who does give randomness its due is the wise-guy stockbroker who sent out eight hundred 'buy' recommendations and eight hundred 'sell' recommendations for security A to 1,600 potential clients. After security A increased in value, he discarded the clients to whom he had sent the 'sell' letters and sent 400 'buy security B' and 400 'sell security B' letters to the others. Security B decreased in value, so he again ignored the first group and sent 200 'buy security C' and 200 'sell security C' letters to the second group. And so on and so on. After 6 such rounds, there remained

25 potential clients. His next letter read: "I've now predicted correctly six times out of six; I was right every time; I suggest you let me manage your money."

The truth is that just as the sequence 111111 is bound to appear in a sufficiently long sequence of random bits, there are bound to be traders that are successful six times in a row within a sufficiently large cohort. This has little to do with skill, Taleb maintains; rather it is more likely to be due purely to probability. At the other end of the spectrum, there are traders who lose money year after year, not because they are dumber than their successful colleagues, but just because the sequence 000000 is also bound to appear in a sufficiently long sequence of random bits. And so, the successful group, full of hubris, attributes their success to their skill, while those in the second group, contrite to a fault, attribute their losses to random bad luck. But luck, be it good or bad, is by definition due to the vagaries of randomness.

Mistaking good luck for skill, flops for bad luck, random noise for meaningful hints, and meaningful signals for random noise are some of the manifestations of cognitive dissonance to which human beings fall victim. Individuals systematically tend to overestimate or underestimate probabilities. In general, people's reasoning about probabilities is not guided by laws of probability, but by rules of thumb. One such rule of thumb is to judge a sequence of numbers by whether it is representative of randomness. Hence 00101101101001 would be considered to be random by human beings, while 01010101010101 would not, even though the appearance of both sequences is equally likely. It is just that the former sequence looks more like what we believe to be a random sequence.

To judge an event or an object by what is typical was called the representativeness bias by the psychologists Daniel Kahneman and Amos Tversky. Other biases to which human beings are prone are the availability bias (judgment based on what comes to mind easily), anchoring (judgments relying on what comes first), confirmation bias (select information that confirms one's preconception), gambler's fallacy (future probabilities depend by past events), and others.

Since people are not good at judging probabilities and at identifying randomness, they substitute *decision weights* for probabilities, as Kahneman and Tversky described in 1979, in their famous paper "Prospect Theory." These decision weights exhibit characteristics that do not follow all laws of probability theory. For example, highly unlikely events are either ignored or overweighed, the difference between high probability and certainty is either neglected or exaggerated, the weights of all possibilities do not add up to 100%.

Of course, such errors in judgment can lead to irrational decisions. The problems are exacerbated because the natural response to cognitive dissonance is to rationalize existing beliefs and to reject information that conflicts with them. No wonder then, that even sophisticated economic agents who believe themselves quite rational and who are well informed about cognitive biases, are still likely to fall victim to them. They are *fooled by randomness*.

≈

When the Scottish botanist Robert Brown (1773–1858) peered through his microscope in the early nineteenth century, to look at the pollens of a certain flower suspended in water, he noticed that they performed strangely random motions in all directions. The cause for the jittery movement, a.k.a. *Brownian motion*, would become known only a century later. It was discovered by no less than Albert Einstein. He surmised that the pollens were being pushed around randomly by the water molecules...though at that time the reality of atoms and molecules was still only conjectured and their existence was definitely established only in the 1950s.

The key word in the previous sentence is 'randomly.' The molecules slam into the pollens from all directions, pushing them this way and that, without any rhyme or reason. In one dimension, that is, along a line, the phenomenon can be visualized by a drunkard who staggers along the sidewalk, randomly taking a step forward or backward. Brownian motion along the sidewalk is called a *one-dimensional random walk* and can be modeled by a sequence of random bits.[1]

Towards the mid-nineteenth century, the French stockbroker Jules Regnault had suggested that stock prices change according to a random walk model but his work remained largely forgotten. Several decades later, a compatriot, the mathematician Louis Bachelier (1870–1946), discovered that Brownian motion governs not only microscopic particles suspended in water but that its one-dimensional counterpart, the random walk, also occurs in the stock market. In his doctoral thesis, Bachelier maintained that since stock prices are just as likely to go up as they are to go down, the average profit that a speculator can expect to make is zero. This insight, and

---

[1]   The most important consequence of a one-dimensional random walk is that the expected distance at which the drunkard finds himself after t steps is $\sqrt{t}$.

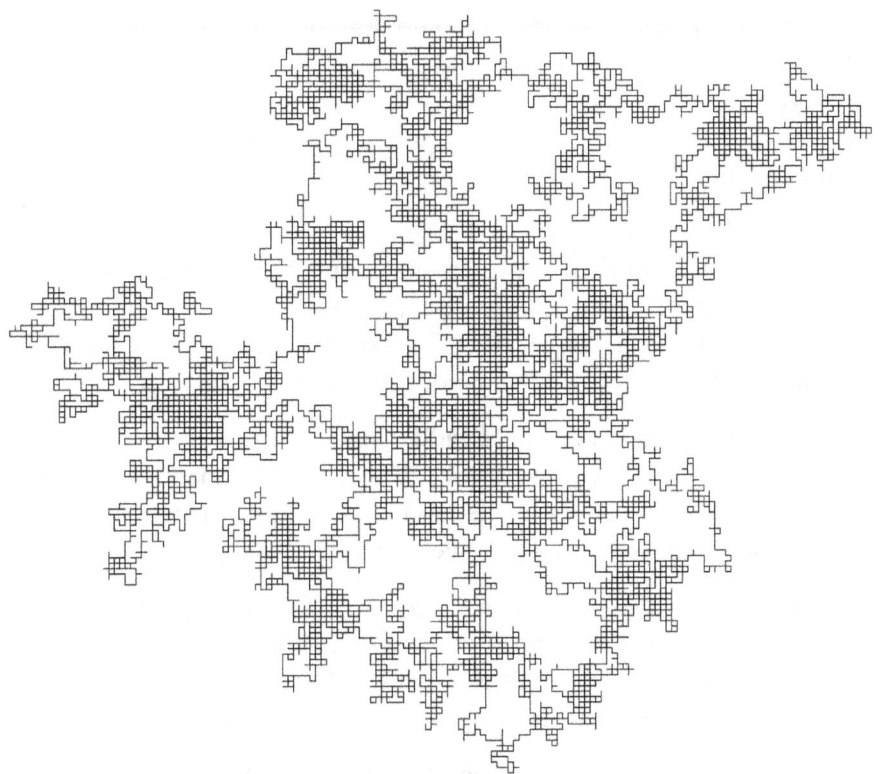

FIGURE 7.1   Two-dimensional random walk, 25,000 steps.

(From https://commons.wikimedia.org/wiki/File:Random_walk_25000_not_anima
ted.svg).

several others, were the basis for the development of the famous Black-Scholes options pricing formula which led to the Nobel Prize in economics in 1973.

Not everybody considers the random walk hypothesis to be valid. Some market players, so-called technical analysts or chartists, believe that if you plot the prices on graph paper, the shape of past movements (e.g., head-and-shoulder pattern, candlestick, inverse head and shoulders, triple bottom reversal, …) determine the future behavior of the stock market. This is humbug, of course.

On the other hand, serious market professionals and academics are convinced that the movement of stock prices is not entirely random; they believe that the stock's true value can be determined on the basis of the so-called fundamentals, macroeconomic indicators like unemployment, inflation, supply and demand ..., and microeconomic indicators like debt-to-equity ratio, price-to-earnings ratio, operating efficiency, inventory levels, ... In a book that may have been one of the first bestsellers in financial theory, the Princeton University economist Burton Malkiel declared that

> Despite its plausibility and scientific appearance, there are three potential flaws in this type of analysis. First, the information and analysis may be incorrect. Second, the security analyst's estimate of 'value' may be faulty. Third, the market may not correct its 'mistake' and the stock price might not converge to its value estimate.

To Malkiel's first point one might add that not only may information and analysis be incorrect but there may be factors that the analysts don't even know about. Recall Donald Rumsfeld's Department of Defense obfuscation masquerading as a news briefing of February 12, 2002.

> ...as we know, there are known knowns; there are things we know we know. We also know there are known unknowns; that is to say we know there are some things we do not know. But there are also unknown unknowns – the ones we don't know we don't know.

Indeed, there are many unknown unknowns which make the world go round...and make stock prices go up and down.

So, the jury is still out: some people believe that fundamentals carry information about future stock prices, while adherents of the random walk hypothesis believe that stock prices move up or down randomly, like coin tosses. I again quote Malkiel:

> No scientific evidence has yet been assembled to indicate that the investment performance of professionally managed portfolios as a group has been any better than that of randomly selected portfolios.

The random walk hypothesis has its detractors also on statistical grounds. Benoît Mandelbrot (1924–2010), the father of fractal geometry, weighed in together with Nassim Taleb on the behavior of stock and grain prices, claiming that the traditional random walk models disregard sharp jumps or discontinuities. "We live in a world primarily driven by random jumps, and tools designed for random walks address the wrong problem" they wrote in the *Financial Times*. They defined 'wild randomness' as an environment in which a single observation or a particular number can impact the total in a disproportionate way. So, random coin-toss models like 'up, down, down, up, down, up, up...' do not capture the behavior of stock markets correctly. Occasional instances of *waaaay* up, and *waaaay* down are more apt descriptions of the movements of stock prices.

To summarize, we know that market players who choose securities by inspecting historical charts are definitely barking up the wrong tree; portfolio managers who base their decisions on market fundamentals do not consistently beat the market either; and adherents of the random walk hypothesis are also out. So, how should one buy securities?

Skeptics like Nassim Taleb would claim that one may as well toss coins to decide which securities to buy. Modern portfolio theory provides a more scientific answer: one should purchase shares of *all* corporations quoted on the stock exchange (according to their weight in the economy) in order to diversify one's holdings. Then the random up and down movements will average themselves out and reduce the risk to its minimum.

In this part of the book we asked what random numbers are good for. The answer was gaming and gambling, polling, random sampling, making money, and even choosing office holders. In later chapters, we shall encounter more applications, like encryption and simulation, as well as advanced techniques in the design of algorithms. But let us now ask this: how do we obtain random numbers? In Part III of the book, we shall discuss some methods and techniques that produce the desired random numbers.

III

Random Numbers
How Do We Produce Them?

CHAPTER 8

Alea Iacta Est

Beyond Coins and Dice

Gambling is, at its purest and most abstract, wagering money on a sequence of random numbers. Before we venture into the digital technologies that have been developed to generate random numbers, let us appreciate both the simplicity and the surprising breadth of the analog predecessors that have been used by players and gamblers throughout the ages. Coins, dice, spinning tops, numbered ping-pong balls, shuffled playing cards, roulette wheels are the archetypes of physical random number generators. How good are they?

The simplest way to produce random numbers is to toss coins. With this, random binary digits are produced, that is, bits of zeros and ones. They are good for answers to yes/no questions or to decide between one of two options, for example which team gets to kick off in a soccer match. To obtain a random number between zero and q, where q is of the form $2^n - 1$, the coin must be tossed n times; at every toss, the new bit is appended to the end of the string and the digits are read from right to left. A string of binary digits would be of the form, say, 1001001 to indicate the decimal number 73.[1]

---

[1] If q is not of the form $2^n - 1$, one simply chooses n sufficiently large and ignores any numbers larger than q.

DOI: 10.1201/9781003641520-12

FIGURE 8.1   Tossing a coin for the kickoff. (From https://commons.wikimedia. org/wiki/File:Heads_or_Tails%3F_(8483623442).jpg).

A paper published in 1986 investigated whether coin tosses produce random bits. Would the coin's final orientation be influenced if the coin starts out with heads up? With an initial height *a*, an initial upward speed *u*, and an angular velocity *w*, the *u* – *w* space is partitioned into regions that define whether the coin would land heads or tails. Depending on how many revolutions the coin performs before landing – albeit on a soft surface so there would be no bouncing – the laws of physics determine the final orientation of the coin. The deeply unsurprising conclusion, formulated and proved as a mathematical theorem, is that the coin lands heads half the times and tails half the times. According to the theorem, this is so, independently of the initial orientation of the coin.

The six-sided cube, the familiar die, is the best-known generator of random numbers between one and six. Die (plural: dice) derives from the Latin word *datum* which indicates 'given' (which itself derives from *dare*, to give).[2] Used in all kinds of games – monopoly, snakes and ladders, backgammon – the number exhibited by the die indicates how many steps the player advances on the board.

---

[2]   Though when Julius Cesar crossed the Rubicon he did not say '*datum iactum est*' to indicate that the die is cast, but '*alea iacta est*'. *Alea* gave rise to *aléatoire* which quite appropriately means 'random' in French.

To distinguish the faces from each other, pips are drilled into them, with the pips on opposite faces adding up to seven (one and six, two and five, three and four). Is the die fair, are the numbers produced by rolling it uniformly distributed and hence truly random? The answer is yes, unless the die is 'loaded,' that is, if one side is heavier than the other ones. Due to gravity, a loaded die would land more often on the heavier face.

To wit, since the face showing 'one' has only one hole drilled into it, while the opposite face has six, the 'one face' is heavier than the 'six face.' Hence the die would often come to rest on the face showing one, and six pips would show up on top. To prevent such bias, the pips drilled into the faces of the die must be filled with material of the same density as the material of which the die is made. To prevent further shenanigans, dice for professional gamblers are produced from transparent material so that all players can ascertain that the material contains no bubbles and that nothing heavy is embedded on one of the sides.

In ancient times, die-makers did not always take care to create fair, symmetrical, cubic dice. In a study of 28 Roman dice from about the first century BCE, excavated in modern-day Netherlands, archaeologists found that most of them were not cubes but cuboids, that is, brick-like shapes, elongated along one axis. Usually, the ones and sixes were placed on the elongated (and opposite) faces and were therefore most likely to show up in tosses. In fact, some of the dice had a 43% chance of falling onto one of the elongated faces instead of the 33% of a perfect cube.

Archaeologists have been puzzling about this asymmetry. While lopsidedness was to be expected in handmade objects dating to antiquity, such high rates suggest that the irregularities were not simply due to chance or random manufacturing errors, especially when post-Roman handmade dice were much more symmetrical. Some scientists speculated that it may have made game play more exciting; or that cheaters with greater experience may have exploited the differing odds, without actually knowing anything about probability theory.

The authors of the study were of a different opinion. To the Ancient Romans the throw of a die was governed by fate and by the favor of the gods, regardless of its shape. That the probabilities differed for the faces on cuboids did not matter, as long as the chances to fall on any of the faces were strictly greater than zero.

※

Just as cubes are dice with six sides, coins are actually nothing more than dice with only two sides. There are many more shapes of dice. Solids that can serve as dice belong to the class of three-dimensional objects called convex polyhedra. Let us investigate them a little bit.

Two-dimensional shapes demarcated by straight lines that connect to form a closed area are called polygons. Three-dimensional solids that are demarcated by flat polygonal faces are called polyhedra. If all lines connecting any two points of the polyhedron lie wholly within the interior of the polyhedron, the solid is convex (i.e., no star shaped objects with pointy thingies protruding). The six-sided conventional die, one of the so-called Platonic solids, is a convex polyhedron. So the question is, do more convex polyhedra exist that can serve as random number generators?

Yes, they exist. First of all, we have the four other Platonic solids: the tetrahedron with 4 faces, the octahedron with 8, the icosahedron with 10, and the dodecahedron with 12. The faces of these five polyhedra are identical – squares for the cube; triangles for the tetrahedron, the octahedron, and the dodecahedron; pentagons for the dodecahedron. Are they fair?

There is also the teetotum, a spinning top with 4, 6, 8, or 12 sides, going back to the Ancient Greeks and Romans. Though not totally symmetric, it is symmetric around the spinning axis. Each side is marked with a letter that indicated whether the player received or lost money. In Jewish tradition, the four-sided version is called the *dreidel*. Though gambling is frowned upon by the rabbis, children are encouraged to play with it on the holiday of Hanukkah.

Mathematically, a die is fair, or unbiased, if it is symmetric with respect to all of its faces: any face can be transformed into any other face by rotation, by reflection, or by a combination of both.[3] The faces can be switched without changing the probabilities of the generated numbers. You could color the faces and decide the red side is one, the blue two, and so on. After several tosses you could exchange the colors: now green indicates one, brown two … . In whichever way you color them, the distribution of numbers must be the same for fair dice. The Platonic solids are symmetric with respect to their faces, vertices, and edges (i.e., each face and edge and vertex is indistinguishable from any other). Hence, they are fair.

Are there additional solids that can be used as dice, even though, unlike the Platonic ones, they are not symmetric? For example, the rhombic triacontahedron, made of 30 rhombi, is symmetric with respect to the faces – any face can be rotated into any other – but it is not totally symmetric

---

[3] The set of all these transformations is called symmetry group.

because at some vertices 5 rhombi meet, at others 3. So, it cannot be considered fair; the bouncing behavior could differ depending on whether it first meets the surface on a vertex where three or where five rhombi meet.

There exists a whole zoo of fair dice. Most are symmetric with respect to their faces, and their faces have the same relationship with the center of gravity (they are called isohedra). The octagonal trapezohedron, for example, with 16 kite-shaped faces, the deltoidal icositetrahedron with 24 kites, the tetrakis hexahedron with 24 isosceles triangles, the triakis icosahedron with 60 isosceles triangles, the pentagonal hexecontahedron with 60 pentagons, the disdyakis triacontahedron with 120 scalene triangles, etc. Persi Diaconis and Joseph Keller catalogued the fair polyhedra, among them are the 5 Platonic solids, the 13 so-called Catalan solids, and 2 infinite classes, the so-called dipyramids and trapezohedra. (Dipyramids are two pyramids symmetrically placed base to base. Since the bases of pyramids can have any number of sides, the class of dipyramids is infinite. Similarly, trapezohedra are bipyramid-like solids but with kite-shaped faces.) Altogether, there are 30 families of isohedra; they have even number of faces between 4 and 120 (except for the dipyramids and trapezohedra, which can have even number of faces between 6 and infinity).

And then there are solids that can be used as generators of random numbers although they are not even symmetric in their faces. As an example, let us look at the British one-pound coin, introduced in March 2017. It has a diameter of 23.43 millimeters and is 2.8 millimeters high. Its most striking feature is that its circumference is not a round circle but a 12-sided polygon (a 'dodecagon'). Like any coin, when tossed, it will fall heads half the time (with the queen's or king's likeness on top), and tails half the time. But now imagine thickening the coin until it is, say 50 centimeters high; it would resemble a long, fat pencil, with heads on one end, tails on the other. If this 12-sided pencil-like cylinder is tossed into the air and falls down, it will most probably not come to land on one of the two faces but will almost surely fall on one of the 12 sides. So, the probability of falling either heads or tails will have dropped from 50% each to near zero for both. Now successively shorten the pencil-like cylinder. By continuity, there must come a point where the probabilities of landing heads or tails or on any one of the 12 sides are identical (and equal to $1/_{14}$). At this point, the pencil/coin/cylinder would be fair! Note that if the pencil had an odd number of sides, one could, by the same construction, create a fair die with an odd number of faces.

One can create more esoteric 'random number generators,' for example by nesting solids inside each other. For example, a 6-sided cube can be embedded in another 6-sided cube that is made out of transparent acrylic, to produce random numbers between 2 and 12 by adding the dots. A disdyakis triacontahedron (the 120-faced isohedron) placed inside another disdyakis triacontahedron would allow random numbers between 2 and 240. Of course, one could simply throw two 6-sided or 120-sided dice consecutively or simultaneously … but the embedded dice are much fun.

❧

In principle, if all initial conditions of a toss are known – velocity, angular velocity, density of the air, elasticity of the die, elasticity of the table, the height from which it is thrown, properties of the table's surface, friction parameters, etc. – the face on which the die will eventually come to rest should be computable. After all, the dynamics of the die (or coin) are determined by Newton's laws of motion. Therefore, if all parameters are established, the outcome of the throw is, in principle, predictable. To investigate whether this is really so, four Polish engineers performed some experiments.

They reasoned that since a die's dynamic is described by perfectly deterministic laws of classical mechanics, the conditions just before the throw (position, configuration, momentum, and angular momentum of the die), in addition to the viscosity of the air, and the friction between the die and the surface on which it lands, should determine the throw's outcome. To verify their claim, the engineers created a three-dimensional model of the die throw and compared the mathematical results with the results of the experiments. They constructed a device that tossed dice from a height of 30 centimeters onto a table. The device allowed them to set the initial conditions very precisely, and a high-speed camera followed the dice, capturing their motions at 1,500 frames per second.

As it turns out, the die's unpredictability derives not only from the rotations in the air but, more significantly, from the dozen or so bounces after it touches ground for the first time. What the authors found was surprising: the die would most often land on the face that was at the bottom at the beginning of the throw.

But the equations that were required to arrive at this conclusion were highly non-linear. This means that the results are extremely sensitive to the initial conditions; a tiny imprecision in the measurements of the initial conditions would distort the result entirely. Hence, in order to make a prediction, initial conditions and parameters would have to be known to many digits after the decimal point and the slightest error would throw off the prediction. This phenomenon is known as chaos theory.[4] It says that if a mathematical model contains non-linear equations, the results are very sensitive to the initial conditions. Even the tiniest measurement errors or imprecisions can build up and give completely different results. The well-known butterfly effect, for example, says that due to the many non-linearities in the equations that model the weather, even the flaps of a butterfly's wings in Australia may influence the weather in Texas.

Having said all this, fair dice produce random numbers only if the person who throws them is honest. Simply giving the die a gentle push and having it slide along the table constitutes cheating. To thwart that, casinos insist that the die hit the perimeter of the table after the throw, and bounce back. The perimeter, studded with foam spikes, makes the die bounce back in a completely unpredictable way, thus frustrating any attempt to control it.

❧

So far, we discussed flat or pointy objects to generate random bits or numbers (coins, cubes, tetrahedra, octohedra…). The more sides the various isohedra have, the more they approach a round ball; the disdyakis triacontahedron (the 120-faced isohedron) gets quite close. And indeed, round objects with symmetry in all directions, not just in two as with a coin, or six as with a six-sided die, are used to generate random numbers in roulette. The randomness of ping-pong balls that twirl around in turbulent air, colliding with each other and with the sides of the container, makes them the preferred implement for modern lotto drawings. Presumably, the balls' very roundness gives the gamblers faith in their unbiased randomness.

How about playing cards? Can picks from a pack of shuffled cards serve as random numbers?[5] To answer this question, we must define a measure for the degree of randomness in a deck of shuffled cards. The key concept here is 'shuffled,' another term for permutated.

A deck of playing cards is said to be randomly arranged – excuse the oxymoron – if the distribution of the arrangements is uniform, that is, any arrangement of cards is equally likely. Since there are $n!$ possibilities of arranging n playing cards (about 8×10^{67} for 52 cards), the probability of each conceivable arrangement is $1/n!$. For 52 cards this corresponds to 1.2×10^{-68}. In order to use playing cards as a source of random numbers, one needs a suitable method of shuffling them, such that the resulting

---

[4] The Dynamical Systems Collective in San Diego had a hand in developing chaos theory.
[5] The tens are identified as zeroes, while jacks, queens, and kings are simply ignored.

distribution of arrangements becomes as close as possible to the uniform distribution.[6]

We will consider the riffle shuffle: cut the deck in two 'semi-decks' and then 'riffle' them in such a manner that the cards of both semi-decks interleave. Do that several times. The question now becomes how many times does one have to riffle shuffle so that the resulting distribution of arrangements approaches uniformity.

Again, it was the Stanford mathematician and magician Persi Diaconis who, together with collaborators, analyzed the situation. With 52 playing cards, they found that one, two, three, or four riffle shuffles do next to nothing to randomize the deck. Only after the fifth shuffle does the deck begin to move in the direction of randomness, slowly at first and then faster.

The mathemagician and his collaborators derived a precise formula: to achieve randomness, the number of shuffles of a deck with n cards must be larger than $^3/_2 \log_2 n$ (for large n). So, for a deck with, say, 1,000 cards, 15 shuffles would be needed to achieve randomness. With a regular deck of cards ($n = 52$), nine would approach perfect randomness, but seven shuffles suffice to get the deck to lie halfway between total order and perfect uniformity.

A different approach to the randomness of playing cards is based on information theory pioneered by Claude Shannon. It regards the act of shuffling to be a process of destruction of information. According to this approach, fewer riffle shuffles are required. After five shuffles, 96.5% of the information that was contained in the ordered deck is destroyed, after six shuffles 99% is gone.

We see that some of these analog devices are better than others at producing random numbers. But in any case, the manipulation of physical objects like coins, dice, wheels, playing cards, ping-pong balls is far too cumbersome for practical use in modern applications. And since even playing and gaming is mostly electronic nowadays – slot machines, digital roulette, online poker, video games, etc. – the required random numbers must be produced with the help of more sophisticated methods. I will describe some such techniques in the next three chapters.

---

[6] For the mathematically minded: to measure the distance of the distribution from uniformity, one uses the so-called L^1 metric on the space of densities.

CHAPTER 9

Simplicity in Motion

Mechanical Random Numbers

In between the times when coins and dice were tossed and today's high-performing digital random number generators (RNGs), a curious variety of devices were invented and patented to generate random numbers. By making use of the chaotic principles which underlie basic physical phenomena, these devices and processes bridge the simplicity of dice and coins with the sophistication of modern RNGs. Let's explore some of these imaginative physical devices.

Over the past three centuries, many thousands of patents have been granted all over the world that contain the term 'random number' in the title or abstract. What I present in this chapter is by no means exhaustive, nor always correct prioritywise. It is more a *tour d'horizon* than a history of patented schemes to create randomness.

The earliest known American patent for the generation of random numbers dates from December 4, 1815. Titled simply 'Lottery,' the patent was granted to Joseph Vannini of Washington, DC, who, a few years later, became the manager of the Virginia Lottery Office in Richmond, VA.

Over the following two decades, 15 more patents were granted, for example, "Calculating and Drawing Lotteries" by J. J. Cohen of Baltimore, MD, in 1823; "System of Lotteries and Mode of Drawing" by F. W. Dana of

DOI: 10.1201/9781003641520-13

Boston, MA, in 1825; or "Instituting and Drawing Lotteries" by J. K. Casey of New York, NY, in 1931. Unfortunately, all records of these patents were lost in a fire in the US Patent Office in 1936, and no details beyond their titles are known.

But we do have some indications about Vannini's patent, because the inventor, by then residing in New York City, NY, filed another patent with the title "Improvement in the Mathematical Operation of Drawing Lottery-Schemes." Dated July 18, 1840, it was granted patent no. 1,700. The improvement was indeed necessary because, as it turned out, his earlier patent was woefully inadequate.

How did this come about? Well, disputes about patent infringements are legion, and Vannini's invention of 1815 was no exception. Lotteries presented a lucrative business opportunity, agencies sprang up who organized them and collected a part of the proceeds. One of the foremost lottery brokers of its time, the firm McIntyre & Yates, acquired the rights to Vannini's lottery scheme. At the same time, the firm Paine & Burgess entered into an agreement to raise money for a school through a lottery. When it became known that the winning tickets of the school's lottery would be chosen according to Vannini's invention, Vannini and McIntyre & Yates filed a lawsuit against Paine & Burgess. After initial wrangling, a preliminary injunction, and a consequent dismissal, the case came before the Delaware Court of Errors and Appeals.

The defendants maintained that Vannini should never have been granted a patent. They argued that his scheme "was not original: that the same plan was to be found in Dobson's *Encyclopaedia*, tit. Lottery." Furthermore, Vannini's patent "is not connected with any construction of matter or application to machinery. It is a mere abstraction. The invention is not new. It is simply the common arithmetical rules of combination and permutation."

As unfavorable as the argument was to the appellants, the judge deemed it unnecessary to express any opinion on the fundamental question. He did not give a hoot about whether Vannini's invention was worthy of a patent or not; he had an entirely different view of the matter. His point was that the very act of organizing a lottery was itself forbidden in the State of Delaware. "We had in force an act of assembly prohibiting lotteries, the preamble of which declares that they are pernicious and destructive to frugality and industry and introductive of idleness and immorality, and against the common good and general welfare." To drive home his point, the judge further thundered that "a person might with as much propriety claim a right to commit murder with an instrument because he held a patent for it as a new and useful invention." This profound legal argument was cited in later cases, for example, when a court held that the production of dynamite may be forbidden next to a schoolhouse, even though the manufacturers had a patent for its production.

Then the judge delivered a bombshell: "Taking all the facts stated in the bill to be true, the plaintiffs. … appear more justly deserving punishment by way of a criminal prosecution, than protection from the extraordinary powers of a Court of Equity."

Tough words indeed; not only were the plaintiffs not going to be indemnified, but they were to be punished. This was all the more bizarre, since Vannini was to be prosecuted simply for proposing "common arithmetical rules of combination and permutation." (Strangely, the judge had no harsh words for Messrs Paine & Burgess who should have been at least as guilty.)

Luckily for the plaintiffs, as far as is known they were spared criminal prosecution. But the allegation that the scheme was just a well-known matter of simple mathematics must have stung Vannini profoundly. So, in 1840, after having gone back to the drawing board, he applied for, and was granted patent no. 1,700 for an improvement of his scheme.

It was again lots of arithmetic and combinatorics: 11,748 tickets were to be printed with 5 single digits, as well as 10 two-digit, and 10 three-digit combinations of numbers between 1 and 90. The tickets would be printed in sets of 18, with each ticket containing, apart from the other number combinations, 5 of the 90 single-digit numbers ($5 \times 18 = 90$). On the day of the drawing, the generation of random numbers was achieved by a five- or six-year-old boy who would pull five lucky numbers from a wheel without replacement. (Note: the actual number generator was the little boy.)

The system was so complex that later experts were at a loss. For one, Vannini had overlooked the fact that the number of sets – that is, the number of tickets (11,748) divided by the number of tickets in each set (18) – is not an integer. Since the scheme was apparently never put into practice, it remains unknown if and how it would have worked.

❧

In 1883, at the Imperial Patent Office in Germany, J. E. Barbé, a Frenchman from Paris, obtained patent no. DE26,471 for an RNG that would, in the inventor's words, offer players much more entertainment than most other hazard games. Fashioned as a pistol, it was not meant for lotteries but as an amusement. A small box inside the barrel contained three straps on which digits between zero and nine were embossed. The straps were connected to a knob outside the barrel. When the player rotated the knob, the straps rotated without the player being able to see them. When he or she then pulled the pistol's trigger the box was projected to the end of the barrel and a three-digit random number appeared. To heighten the suspense, a small explosion could be added to the device. *Voilà*, endless fun.... .

Patent no. DE129,735 was granted by the German Imperial Patent Office to George McMullen and Joseph Charles, two inventors from Perth, Australia, in 1901. The contraption consisted of a box containing four disks with the digits zero to nine printed on their peripheries. Independent belts attached each disk to a motor. Once the electricity was turned on, the belts would set the four disks into motion and it was up to the player to decide when to shut off the current. At the moment that the disks come to a stop, four digits would be visible through a window of the box, thus creating the four-digit random number. By the way, the inventors remarked there was no need to put numbers on the disks; there could just as well be symbols or images.

Remarkably, the two astute inventors added an interesting twist to their contraption which indicated that they were quite aware that the resulting numbers were not completely random. After all, a nimble-fingered player could possibly influence the appearance of the supposedly random digits by astute timing of the on/off switch. To make cheating more difficult, the two inventors suggested that the disks have gears with different diameters, thus letting them rotate at different speeds, and to have the straps cross themselves, so that some wheels rotate in opposite directions. *Bei dieser Einrichtung wird die Vorausbestimmung bezw. Berechnung der ... sichtbaren Zahl fast vollkommen unmöglich gemacht.* (This feature renders the prediction response, the computation of the visible number almost completely impossible.) Note the clause *fast vollkommen* (almost completely). The inventors knew that the result was not completely random. Arguably, Patent no. DE129,735 was not only the prototype of the one-armed bandit of modern casinos but could be said to be one of the first generators of *pseudo*-random numbers.

In 1928, the French *Ministère du Commerce et de l'Industrie* (ministry of commerce and industry) granted patent no. 646,761 to Tomozo Nose from Japan for an apparatus that picks balls for elections and the like. A vertical drum is rotated around its axis by a handle, with the balls – the names of the candidates inscribed on them – twirling inside. One ball is allowed to escape and the person whose name the ball bears is elected. Of course, the balls could be numbered from zero to nine, thus producing a random digit each time, on condition, of course, that the escaped ball is put back into the drum.

In 1956, the German Patent Office, by then no longer of Imperial imprimatur, granted one to Josef Rebl, patent no. DE1738167U, for a device to generate random numbers for board games between one and six with – no surprise there – dice. But, as Rebl explained, dice have the annoying habit of rolling off the edge of the table, much to the displeasure of the players. Tossing dice in a beaker is also no panacea, he wrote, because the vigorous thump when the beaker is placed upside down on the table by enthusiastic players often causes game pieces to fall over. So Rebl designed a transparent bell-shaped top to be placed in the middle of the table. Players on all sides could activate a lever which agitated the bottom of the device and made the die jump up and down and spin around without danger of falling off the table. An added advantage, Rebl remarked, is the device's hygiene since no germs are passed from someone's hand to the die, and from there to someone else's hand.

Back to the United States, as the drawing of lottery numbers on television went prime-time, blowing air into containers with ping-pong balls became the method of choice to generate the winning random numbers. The reason for its prominence is that turbulence, described by Richard Feynman, the Nobel Prize winning physicist, as the most important unsolved problem in classical physics, is probably the most easily visible physical process exhibiting randomness.

In 1936, Harry Rochwarg from New York applied for a patent for an invention that he simply named 'Amusement Apparatus.' The patent was granted a year later and given the number 2,091,883. The invention consists of a transparent bowl with an inlet in the middle of the floor, connected to an air blower, and an outlet in the middle of the top, connected to two separate receptacles. Several balls marked with numbers, most likely ping-pong balls, were to be placed on the bowl's floor. With the help of a lever the player could cause the floor to tremble and make the balls spin around. The objective was to direct a chosen ball towards the air inlet and at the precise moment send a puff of air which would make the ball rise and exit through the opening at the top. The strength of the puff, controlled by the player by a second lever, determined into which of the two receptacles the ball would fall. The game's objective could be, for example, to direct odd numbers in rising order to the left, even numbers in rising order to the right. The Amusement Apparatus could be coin operated and the available time to direct the marked balls towards the appropriate receptacles could be limited to, say, a minute.

The turbulence caused by the air puff is quite random, of course, but in Rochwarg's apparatus this could be overcome by the player's skill in operating the two levers. Hence, the Amusement Apparatus had little to do with random numbers. But it was a forerunner of what would half a century later become the preferred means of choosing random numbers at raffles, in bingo parlors and at TV shows.

© Chat GPT

The cudgel was taken up by Richard B. Dunnigan of Washington, MI. His patent no. 4,786,056 of November 1988 used the air-blowing idea of Rochwarg's Amusement Apparatus, not to amuse players, but with the express aim of manufacturing a "random number generator assembly for mixing and randomly selecting balls indicating events or numbers."

It was the time when personal computers were just about to become ubiquitous, and Dunnigan was well aware that "random numbers generated by a properly programmed computer may be theoretically capable of approaching a statistically perfect random number mix." But, he cautioned, the black-box approach to the generation of random numbers "is unsuitable for most games in that the participants cannot see or oftentimes understand the process by which the numbers are generated and may therefore lack confidence in the random number selection."

In order to avoid black-box-generated random numbers, he developed a mechanical generator of true random numbers, based on Rochwarg's Amusement Apparatus. His device consisted of a transparent mixing chamber, into which balls, marked with digits, were placed. The chamber had a perforated floor which sat above a fan motor; one of the mixing chamber's walls contained an aperture just large enough that one ball could escape through it at a time.

An operator would turn on the motor's power switch which activated the blower. The passage of air through the perforated floor would create turbulence in the mixing chamber and the balls would twirl around until one of them escaped. "The passage of a ball through this aperture is considered the generation of a random number."

Another patent that I picked at random (pun intended), Patent no. 4,786,056, is only one of many more air-blowing lottery machines that preceded and followed Dunnigan's creation. Not all made their inventors rich. Patent no. 4,786,056 expired in 1993 due to a failure to pay the maintenance fee to the patent office.

Most air-blowing machines can be said to generate random numbers. Note, however, that they do not generate random sequences! Once a ball has been ejected, its number cannot appear again. Whenever a true, unbiased random number sequence is desired, each ejected ball must be replaced before the next twirling of the balls.

❧

This chapter covered patented methods that involve physical manipulations of items – pulling a trigger, blowing ping-pong balls – thereby highlighting simple, mechanical processes that have been, and still are, used to generate random numbers. The randomness derives from discrete events, such as the positions or order of items that have been thrown, tossed, shuffled, or moved by air currents. While schemes like the ones described are suitable for board games, games of chance, and amusements, they are much too slow for most computer applications. These require hundreds of thousands or even millions of random numbers within seconds.

CHAPTER 10

Complexity from Nature

Chaotic Random Numbers

Remember how the RAND Corporation used an electronic device, in the late 1940s, to generate one million random digits? A gas-discharge voltage regulator tube – a common tube widely used in regulated power supplies at the time – was the source of the random pulses that were passed to a digital counter. But the result was underwhelming: it was not to everybody's taste to leaf through a 600-page book in order to choose a 5-digit random number. And anyway, a million random digits (i.e., 200,000 five-digit numbers) is peanuts for simulations that may require many times that amount. One would think that with the advent of computing, the generation of random numbers on mainframes, or even on one's own PC or laptop should became possible. Well, as it turns out, it's complicated…

With computers being by definition deterministic, true random number generators cannot rely on them. Only physical phenomena can provide genuine randomness. For real-world purposes, it is coins, dice, floating ping-pong balls, and the like that are able to generate truly random bits and numbers. However, these implements are far too slow to be of any practical use, except for games and lotteries. Fortunately, various natural phenomena, such as the sound of radio static, the output of a noisy diode, the motion of clouds, the thermal noise given off by resistors come to the rescue; they can serve as sources for true random numbers.

DOI: 10.1201/9781003641520-14

One of the early providers of true random numbers was the corporation Silicon Graphics Inc. of Mountain View, CA. Patent no. 5,732,138, granted in 1998, described a "Method for Seeding a Pseudo-Random Number Generator." The contraption provided numbers that were truly random, fast to generate, bountiful…. and very cool. The inventors, Landon Noll, who held several world records for discovering the largest prime numbers, and two co-workers based their invention on, of all things, LavaLamps. It should be noted that the patent only described how to produce a random seed, that is, the initial first number; the sequence of random numbers that follow the seed would then be generated by a deterministic computer algorithm.

A LavaLamp consists of a glass cylinder that contains blobs made of wax and other materials, suspended in water. A lamp at the bottom, lights up the cylinder's content. As the bulb warms up, the wax expands, becoming less dense than the water, and the blobs rise to the top. As the blobs cool, they shrink, becoming denser than the water, and sink. Then the cycle starts over again.

The beauty of LavaLamps lies in the chaotic, unpredictable movement of the blobs. And this is exactly what the Silicon Graphics engineers were looking for. The blob's chaotic movements are captured on a digital camera and the pixels converted into a sequence of bits. Then it is sent back to the computer. The sequence of truly random bits is used to seed an algorithm that produces the desired series of pseudo-random numbers. Since the seed is truly random, the series is unpredictable and may therefore be considered random.

There's an epilogue to the story. Why bother at all with lava lamps? Noll and his colleague Simon Cooper decided to ditch the lamps, put a cap on the camera's lens and take pictures of nearly complete darkness, punctuated only by background noise. (The snow that one sees on analog TV sets, tuned in between stations.) The pixels of the noise were converted into the random numbers. As a service to the community, Noll and Simon did not patent this invention but placed it in the public domain.

While the engineers at Silicon Graphics were busy in California with their LavaLamps, across the Atlantic, in Ireland, the PhD student Mads Haahr and two of his friends were tinkering with a program to be used for online gambling. They realized early on that any serious gambling engine required

FIGURE 10.1   Lava lamps (From https://commons.wikimedia.org/wiki/File:Lava_lamps_(16136876840).jpg).

proper random numbers, but they were not about to place a radioactive source in their office as the RAND engineers had done. It occurred to them that static noise, picked up by a radio receiver, could be a cheap and elegant way to gather the randomness needed for the generation of true random numbers.

Unfortunately, radio sets at the time had a filter built in that squelched the unwanted atmospheric noise on frequencies between stations. While this was considered a bonus feature by almost everyone, it was not what Haahr and his friends had in mind. Luckily, they found a cheap radio set that did not possess that feature and off they went. They hooked the radio to a workstation, created the website RANDOM.ORG and went live in 1998.

Over the years, Haahr upgraded the setup several times before converting the project into a commercial enterprise. Incorporated in 2010 as 'Randomness and Integrity Services Limited,' the company offers many free and some paid services that involve random numbers. For example, it organizes raffles, sweepstakes, promotional giveaways, and lotteries, acting as an unbiased third party that conducts the drawings in a manner that is guaranteed to be fair and truly random. Though the company now uses modern hardware and software, the random numbers are still generated from atmospheric noise.

≈

The second half of the 1990s were the heyday for generators of true random numbers based on truly random natural phenomena. From California, via Ireland, we now move to the mountains of Switzerland.

John Walker had been the co-developer in the 1980s of the immensely successful drafting software AutoCAD, used by architects, engineers, designers, and other professionals all over the world. To distribute the software, he founded and led *Autodesk*, a company that went public in 1985. Just one year later, he decided that he was not made out to be a manager and resigned from his positions as chairman and president. He remained with *Autodesk* for another five years, albeit as a programmer, but then got fed up with Silicon Valley altogether. By then a multi-millionaire several times over, he decamped, at age 41, to Zurich in Switzerland. Now, decades later, he is still in the Alpine country, in a little village in the French-speaking part of Switzerland, where he maintains his website *Fourmilab.ch*.[1] Far from the hustle of Silicon Valley, he has the leisure to indulge in his many interests, among them astronomy, photography, software, and voracious reading. One of his enduring endeavors is *HotBits*, the generator of true random numbers.

In his quest to use a natural phenomenon to generate random numbers, Walker turned to quantum mechanics. The theory behind *HotBit's* random number generator is that quantum physics is inherently probabilistic.

The theoretical foundations of *HotBits* go back to 1899, when Sir Ernst Rutherford discovered that radioactive atoms decay. For example, the radioactive atom Cesium-137 decays by changing one of its neutrons into a proton and simultaneously spitting out an electron (thereby abiding by the law of conservation of charge). With this, the Cesium atom turns into a stable Barium-137 atom. Though the moment when this occurs cannot be predicted, it is known that the so-called half-life of Cesium atoms is 30 years; this means that one half of the Cesium atoms decay into Barium atoms within that period. Crucially however, for an individual Cesium atom, the timing is totally random.

Walker's brain wave was to employ this randomness for *HotBits*. A Geiger counter attached to a computer detects and measures the time that elapses

---

[1] *Fourmilab* is a word play on Fermilab, the particle physics and accelerator laboratory near Chicago, and the word *fourmi*, French for ants, an early interest of Walker's. Sadly, when I wanted to ask John Walker for comments about the draft of this book, I found out that he had passed away unexpectedly in February 2024.

between a pair of two successive decays (T_1) and between the next pair of decays (T_2). If T_1 is greater than T_2, a zero bit is generated, otherwise a one bit.

Fourmilab subjected the number sequences generated by *HotBits* to a battery of tests, among them the test suite of the National Institute of Standards and Technology; the sequences passed all of them with flying colors. But *HotBits* is not very fast, only about 800 bits are generated per second. To guarantee a sufficient supply, Fourmilab keeps on hand a large inventory of random numbers which is tapped whenever a request comes in. The sequence is sent in secure form to the requestor and then immediately discarded from the inventory, never to be used again.

❧

A two-hour drive from John Walker's village is the city of Geneva, seat of the UN and of CERN,[2] the world's largest particle collider. It is also where the company *ID Quantique* took the production of true random numbers to an entirely new level.

The initiator of the project was a research fellow in the Group of Applied Physics of the University of Geneva, Grégoire Ribordy. As a PhD student in physics, he got interested in quantum cryptography and was lucky enough to have Nicolas Gisin as his advisor. A renowned professor of quantum mechanics, he had gained much experience in applied physics as a communications engineer before he became an academic. Gisin supported Ribordy's theoretical work and they eventually founded the company *ID Quantique* to commercialize their ideas.

Again, *ID Quantique*'s random number generator depends on the inherently probabilistic nature of quantum physics. Take photons, for example, the particles that carry light. Their behavior cannot be predicted; in fact, one is not even sure whether they are particles or waves. Let's stipulate that a particular photon thinks it is a particle. Before its position is observed, it doesn't exist anywhere in particular. Or rather, it exists everywhere with a certain probability distribution. Only after its position is observed, is the specific path that *it was going to take* retroactively determined. It made its decision, where to go, only after it has already gone there.[3]

As described in US patent number 7,519,641, a beam of photons is directed towards a semi-transparent mirror that is placed at 45 degrees to the source

---

[2]   *Centre Européen de Recherche Nucléaire* (European nuclear research center).
[3]   Does this not remind of the retrocausal-psychokinesis humbug of Chapter 3? No wonder, quantum mechanics is confusing.

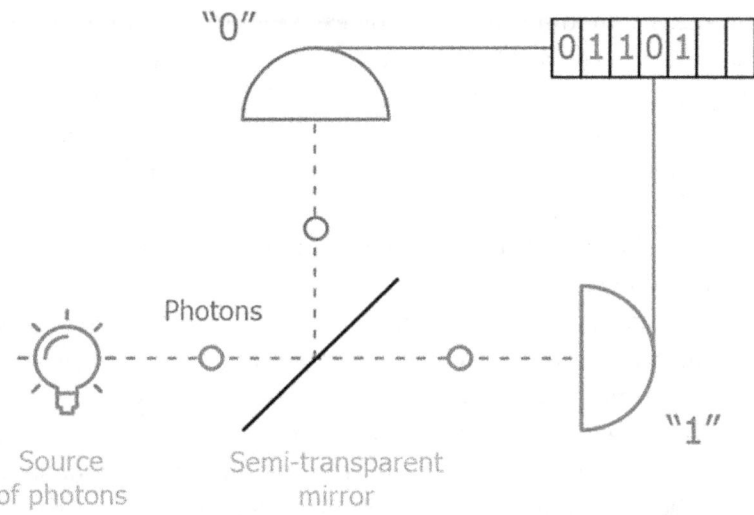

FIGURE 10.2   Optical system used to generate random numbers.© ID Quantique.

of the beam. About half of the photons will pass right through the mirror and hit the detector behind the mirror. The other half will be deflected and hit the detector on the side of the mirror. Which path this, the next, or any future photon will take is completely unpredictable. Whenever a photon hits the detector behind the mirror, one bit is registered, otherwise a zero bit. The quantum process behind this setup provides instantaneous and inexhaustible randomness. The randomness of the numbers stems from the detection probability of the photons in the beam. *Voilà*, a generator of true random numbers.

One would think that this setup requires an entire laboratory and lots of floor space. Surprisingly, ID Quantique managed to pack the beam-projector, the mirror, and the photon counter into a device that measures only 5 × 3.5 × 0.8 centimeters. Small enough to be plugged directly onto the circuit board or inserted into the appropriate slot of a laptop or PC, it is as if a miniature quantum laboratory had been incorporated into one's computer. The device, marketed under the name Quantis, can generate 4 million zeros and ones per second. Even after three quarters of the generated bits have fallen victim to various kinds of quality control, about 15 million digits between 0 and 9 can be generated every minute.

≈

But why look far afield, to lava lamps, atmospheric noise, quantum phenomena, when randomness can be found close by…on your very desk. In spite of computers being deterministic, the hardware itself can serve as a source of randomness. Take, for example, the position of a mouse that has just been moved. Of course, one random position of the mouse on the screen would not suffice since a resolution of, say, 1,920 × 1,080 pixels would provide only about 2 million possibilities. But several mouse positions, taken at intervals, can be used in conjunction to create random numbers.

People's keyboard dynamics – detailed information on the timing when they press and release certain keys – could conceivably be used as sources for random numbers. This technique poses its own problems, however. When Morse code was the means of long-distance electronic communication, the timing between the letters' dots and dashes gave sufficient information on who the sender was. During World War II, the so-called fist of the sender was used to identify the transmitting person. Shifted to modern times, a certain typist's keyboard dynamics could be analyzed by an attacker and the random numbers based on these keystrokes could then be predicted or at least narrowed down.

On the other hand, one may record the exact time when the user hits the 'enter' key in order to create a random number between 0 and 9. Since the internal clocks of personal computers and laptops measure time routinely in microseconds (and even in nanoseconds), the time in the late morning may be stated as 9:50:52.287. The first digits are disregarded because the approximate time of hitting the 'enter' key could be predicted by an attacker or eavesdropper; but the least significant digits (i.e., the last one or two digits) are beyond any user's control and may be regarded as truly random.

Another unpredictable source of randomness is financial data, for example, the closing prices of stock markets, the jumps in commodity prices, etc. There is a caveat, however: Benford's law states that the leading digits of entries in very many data sets are not uniformly distributed. In fact, almost 30% of all datasets (stock prices, populations, street addresses, surface areas of rivers, molecular weights, electricity bills, prices…) is a 1, and only 5% have a leading 9. Hence, again, the leading digits of market data are inadmissible as candidates for random numbers. But even the last digits are not uniform, since they are often rounded up or down to the closest 0 or 5.

With appropriate safeguards, mouse and keyboard methods, hardware noise, financial data may be reasonable sources of randomness. But for serious work – large-scale simulations of high-energy physics, for example – they are too slow to yield useful strings of random numbers.

Websites that pretend to provide truly random numbers have proliferated. However, as one of these sites warns, "random numbers available over the internet and from parties not specifically known to and trusted by the user should not be used cryptographically." To be prudent, this caveat should be extended also to simulations, gaming, polling …

CHAPTER 11

Spooky or Random?

Quantum Random Number Generators

While one may argue about whether tosses of coins, throws of dice, rolls of roulette balls, twirlings of ping-pong balls are truly unbiased, unpredictable, uniformly distributed, most scientists agree – Albert Einstein's reservations ("God does not play dice") notwithstanding – that quantum phenomena are honest-to-goodness random.

And this is where quantum computers enter the picture. These new-fangled thingies, subjects of intense development by billion-dollar corporations, use the properties of quantum physics to perform what classical computers do, and more, but at warp-speeds.

Instead of using binary digits (bits) like the classical computers, quantum computers use quantum bits (qubits) to store information and perform computations. While a bit can be pictured as a light switch that is either off (zero) or on (one), the states of the qubit are, for example, the spin of an electron (up or down) or the orientation of a photon (horizontal or vertical). But in stark contrast to a binary digit which can only be either zero or one, a qubit can be in a superposition of both states. As long as nobody is looking, the electron's spin is both up and down simultaneously; the photon is polarized both horizontally and vertically. Only when a measurement is made, is one of the two states observed ... randomly. So, the

DOI: 10.1201/9781003641520-15

qubit holds two bits of information (both up and down), and n qubits hold 2^n bits of information.

Strange...but that's not all. There is another phenomenon, this time among pairs of qubits. In a letter to Albert Einstein, the German physicist Erwin Schrödinger (1887–1961) called it 'entanglement' (*Verschränkung* in German). Whenever two electrons or photons are entangled and the state of the one is changed, the state of the other will immediately change too, even if they are separated by a large distance. If a scientist changes the spin of an entangled electron in New York from 'up' to 'down,' its counterpart in California will instantaneously change its spin too. This phenomenon is so counterintuitive that Einstein famously called it "spooky action at a distance." (But then again, to Newton's contemporaries, gravitation must certainly also have seemed a spooky action at a distance.)

For quantum computers, superposition and entanglement have vast implications. While the processing power of a conventional computer is doubled when the number of bits is doubled, adding extra qubits to a quantum machine produces an exponential increase in its computational ability. A quantum computer with several entangled qubits in superposition can inspect a vast number of possibilities at the same time. The final result of a calculation emerges only when the qubits are inspected. At that moment, their quantum states collapse to either 1 or 0.

In order to construct quantum computers, huge engineering hurdles must be overcome because qubits are very fragile; the smallest disturbance can throw them out of their superposition and entanglement state. And unfortunately, disturbances are inevitable. Vibrations of nearby atoms, changes in magnetic or electric fields, or stray photons interfere with the qubits' quantum states and cause information to be lost. If the quantum computer could be perfectly isolated from its environment, the unwanted effects of such interactions, called decoherence, could be avoided. But in that case, one would be unable to interact with the computer.

The best one can do is to reduce as much as possible disturbances, like the jitters from nearby atoms. Since vibration is a measure of temperature, the colder the device, the less the jitters, and the greater the chance that the qubits' states of superposition and entanglement can be maintained. The solution is to place the quantum computer into an ultra-high-vacuum chamber, supercooled to a temperature of close to the absolute zero (below −273°C).

Not all quantum phenomena require temperatures close to what reigns in deep space. As described previously, the Swiss firm *IDQuantique*

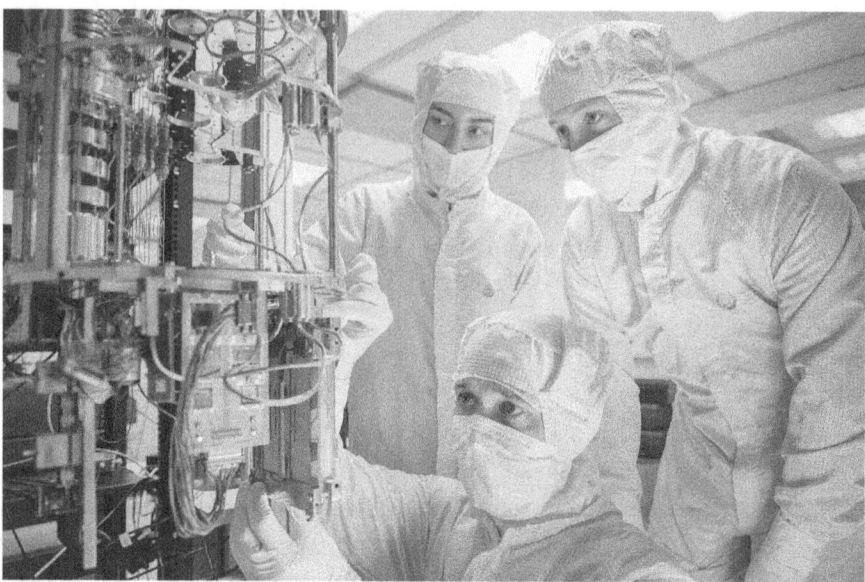

FIGURE 11.1 A team at the FMN Laboratory (at Moscow State Technical University) assembling the cryogenic part of the quantum computer, which provides cooling of superconducting processors to almost absolute zero.

(From https://commons.wikimedia.org/wiki/File:Measuring_a_qubit_leaves_no_room_fo r_error.jpg).

makes use of quantum optics to generate random bits, by splitting a beam of light, such that single photons either transverse, or are deflected by, a semi-transparent mirror. The chip that generates random numbers can be operated at room temperature.

Generating random numbers is not foremost on quantum scientists' minds when they go about their business. There are more significant tasks like weather forecasting, drug development, traffic optimization, and more industries to deal with, like artificial intelligence, healthcare, finance, auto-mation, aerospace, cybersecurity, climate, and many others. But according to Scott Aaronson, a professor of theoretical computer science at the University of Texas, the generation of random numbers – as a proof of con-cept, or simply to show off the technology's capabilities – may become one of the first real-life implementation "not because it's the most important application of quantum computers – I think it's far from that – rather, because it looks like probably the first application of quantum computers that will be technologically feasible to implement."

But are numbers generated by quantum computers truly random? There is the vexatious question of hidden variables: maybe variables exist that control the quantum phenomena deterministically and we simply do not know about them? In that case, the supposedly random numbers would not be random at all. So, the question is whether entanglement, as envisioned by quantum physicists, is real or whether the behavior of supposedly entangled particles is due to the consequences of hidden variables.

As I noted earlier, Einstein remained skeptical. He did not think that quantum mechanics was wrong, just that it was not the whole story. The revered physicist expressed his doubt in a paper written with two post-docs, Boris Podolsky and Nathan Rosen. Titled "Can Quantum Mechanical Description of Physical Reality Be Considered Complete?," it purported to prove that spooky action at a distance was not the complete explanation. The paper turned out to be quite a bombshell. Even the *New York Times* reported, to Einstein's great displeasure, that he and his collaborators, henceforth abbreviated as EPR, believed that "quantum theory, with which science predicts with some success inter-atomic happenings, does not meet the requirements for a satisfactory physical theory."

There are three principles of which the theory needs to take account. First, according to Heisenberg's uncertainty principle, certain pairs of physical quantities, say position and momentum, cannot both be measured precisely at the same time. If, say, a particle's position is measured exactly, its momentum can be determined only within a certain range because the act of measuring its location disturbs its momentum.[1] Second, two entangled particles will have the same momentum even if they have been physically separated. Third, information cannot travel faster than the speed of light.

EPR's contention was that the three principles cannot all hold. To show this, they conceived of the following thought experiment. Two entangled particles fly away from each other at a certain speed. By Heisenberg's uncertainty principle, one can determine either the position or the momentum of each of the particles, but not both. So, let us first measure particle *A*'s position exactly. Now, since we know the speed at which particle *B* travels away from *A*, we can compute particle *B*'s precise location without directly

---

[1] One might picture a blindfolded person trying to determine the position and momentum of a ball rolling on a billiards table. The person can use fingers to determine the ball's location but by touching the ball, he or she would disturb the ball's momentum.

measuring it, and thus without disturbing its motion. Next, we measure particle B's momentum. We can do that precisely because we have not disturbed B's motion. Since B is entangled with A, we automatically also know A's momentum. So, for that very instance, EPR claim that we would know both the exact locations and the exact momenta of particles A and B.

But that contradicts Heisenberg's uncertainty principle which says that we cannot know precisely both the location and the momentum of a particle. Or it would mean that the information about the momentum would have to travel from B to A instantaneously, that is, faster than the speed of light.

All this seemed quite paradoxical. EPR concluded that entanglement is unable to explain quantum phenomena. According to them, the only way to solve the paradox was to posit that hidden variables exist that control both particles simultaneously. The implication is, of course, that quantum phenomena are not random but deterministic and – to jump ahead half a century – that quantum computers could not produce truly random numbers.

Many contemporaries did not believe in the existence of hidden variables. But how could they prove their absence? It is difficult to prove a negative. Well, in 1964, the Irish physicist John Bell (1928–1990) managed to show how it could be done, nevertheless.

Electrons have a spin, up or down and a pair of entangled electrons have opposing spins. They maintain their spin direction even if they are separated by a large distance. Bell envisioned a thought experiment in which two supposedly entangled electrons are sent on their separate ways. After a while, the orientations of their spins are measured. If hidden variables control the orientations, and both spins are measured along the same axis, they will be correlated. But if they are measured along different axes – and even if hidden variables determine them – the spins will only be correlated a certain percentage of times. Bell computed an upper limit to the correlation that can occur due to hidden variables. If the correlation is higher than that, hidden variables cannot be the reason…spooky entanglement must be the answer.

In the decades that followed, many real experiments followed Bell's thought experiment. Without exception, the correlations were higher than what Bell's inequality allowed for the case of hidden variables, thus confirming the theory of quantum physics. Bell himself was actually disappointed; he had found Einstein's skepticism more appealing. But there is no arguing with facts: hidden variables were out, randomness was in.

A year after Bell's paper appeared, Simon Kochen and Ernst Specker published similar results. (I mention this especially because Specker, an extremely kind and humorous man, was the professor who delivered the lectures on linear algebra at the ETH in 1968, when I was a first-year student there.)

※

I already described two random number generators that are based on quantum theory, *HotBits* and *IDQuantique*. Recall that in the former, a Geiger counter detects and measures the time that elapses between a pair of two successive decays of the radioactive atom Cesium-137 (T_1) and between the next pair of decays (T_2). If T_1 is greater than T_2, a zero bit is generated, otherwise a one bit. Since the exact moment when the atom decays is quite unpredictable, the sequence of bits is random. In the latter random number generator, a beam of photons is directed towards a semi-transparent mirror that is placed at 45° to the source of the beam. About half of the photons pass right through the mirror and hit the detector behind the mirror, the other half will be deflected towards the detector on the side of the mirror. Depending on which detector the photons hit, a one bit or a zero bit is registered. The paths that the individual photons take are unpredictable, hence the resulting numbers are completely random.

Are they really? In theory yes, but in practice, no: for the bits to be truly random, the hardware setup would have to be infinitely precise; any minute imperfection could introduce biases, making the bits less random. For example, the mirror would have to be placed at exactly 45.00000...° to the source of the photon beam for the bits to be completely unbiased.

HotBits' online service generates a modest 800 random bits per second; *IDQuantique's* computer chips produce up to a whopping 240 million random bits per second. But even this number is dwarfed by recent advances. As of this writing, the world record in the generation of truly random bits is a million times faster. A team at the department of applied physics at Yale University, composed of scientists from the United States, France, Singapore, Ireland, and the United Kingdom, constructed a true random number generator based on quantum theory, that is six orders of magnitude faster.

Their plan was to capture quantum randomness by exploiting the fluctuations that photons exhibit when emitted by a laser. The material they used for their laser was a translucent semiconductor, about 0.1 millimeters wide, with curved walls. Photons bounced several times between the walls before exiting as a scattered beam. An ultrafast camera captured the light intensities on 127 channels at two terabits per second, thus producing 250 trillion random bits every second.[2]

❧

Part III has taken us through various processes of generating random numbers, from traditional methods like coins and dice to mechanical systems, natural chaos, and quantum mechanics. In Part IV, we will highlight the importance of random numbers in various fields, by exploring the essential roles that they play in cryptography, Monte Carlo simulations, zero-knowledge proofs, randomized algorithms, and the process of derandomization.

---

[2] Translated into alphanumeric characters, this corresponds to about 40 million fat books per second. See also Chapter 13.

IV

**Random Numbers
Why Do We Need Them?**

CHAPTER 12

A Useless Branch of Pure Mathematics

Cryptography

In 1915, in a lecture about prime numbers, the eminent number theorist G. H. Hardy reportedly said, "The Theory of Numbers has always been regarded as one of the most obviously useless branches of pure mathematics." He could not have been more wrong!

The most important, pervasive, ubiquitous application of number theory is in cryptography. Much of the internet, like online trading, exchange of private messages, maintenance of databases, would be inconceivable without encryption. And encryption is currently unthinkable without prime numbers. And prime numbers must, of course, be chosen randomly. We will discuss two methods of encryption. The first, which serves to create a common, secret key, relies on the difficulty of finding the so-called discrete logarithm. The other, which serves to send secret messages, relies on the difficulty to factor large numbers into their components.

≈

The first secret to encryption is that while it is easy to raise numbers to powers, it is much harder to extract roots. Raising 15 to the seventh power is fairly easy, even with just pencil and paper, though it may take a while. Extracting the seventh root of 170,859,375, on the other hand, is much more difficult. And the problem becomes more difficult still, when one uses not everyday arithmetic, but so-called modular arithmetic.

DOI: 10.1201/9781003641520-17

Modular arithmetic 'to base b' says, in short, that the results of computations are reduced by as many multiples of b as possible. To illustrate, let's take 50 as the base b and compute 5^4 mod(50). In regular arithmetic, $5^4 = 625$. To compute 5^4 mod(50), one deducts 50 as often as possible from the result, in our case 12 times ($12 \times 50 = 600$). The remainder is the result: 5^4 mod(50) = 25. This is easy. To find out which root to extract of 25 mod(50) of 25 in order to obtain 5 is much more difficult. This is the secret behind the algorithm which we shall describe now.

Let us say that Alice and Bob want to generate and agree on a secret number. It will be the codeword Bob must state at the entrance to Alice's restaurant in order to gain entry, or the key that each of them can use on their own to open a safe, or the PIN (personal identification number) required to access a bank account. In 1976, Whitfield Diffie, an eternal PhD student, and his electrical engineering professor Martin Hellman devised a scheme to do just that, building on previous work by the computer scientist Ralph Merkle.

The scheme involves modular arithmetic and randomly chosen prime numbers. It became known as the Diffie–Hellman key exchange, which is a bit of a misnomer because it evokes the idea of an existing key being exchanged between two parties; actually, the two parties do not exchange a common key but *generate* one. Another incongruity, this time historical, is that the English mathematician Malcolm Williamson of the British Government Communications Headquarters (GCHQ) had discovered a version of the key exchange method a year before Diffie and Hellman, but was never able to publish his findings because the agency decided that his work would remain classified.

In the following, we present an example of the Diffie–Hellman key generation. If you want to skip the technical details, just continue from the paragraph starting with 'Why does it work?.'

✍

First, we randomly choose a prime number P; we will make do with the number 37, though in practice it will be much, much larger. We call P the *modulus*. Next, we pick a number b, called the *base*, that is smaller than P. We choose 5. Both these numbers, 37 and 5, are published for everybody to see.

Alice has secretly chosen a random number smaller than 37 that need not be prime but has no common divisors with 37 (except 1), say 12, and encrypts it in the following manner: 5^{12} mod(37). The result is 10, which Alice sends to Bob over an open channel. Bob also has a secretly chosen random number smaller than 37 that has no common divisors with 37, say 17, encrypts it, 5^{17} mod(37) = 22, and sends his result over an open channel to Alice.

Alice decrypts the result with her secret number: 22^{12} mod(37) = 26. At the same time, Bob decrypts Alice's message with his own secret number. Lo and behold, it is the very same number: 10^{17} mod(37) = 26. This is the key which Alice and Bob now share.

The reason that this scheme works is because when raising a number to a power, and then raising it again, the exponents can be interchanged. For example $(5^7)^3 = (5^3)^7$. In modulus operations, it is the two secret numbers that can be interchanged:

$$\left(5^{17} \bmod(37)\right)^{12} \bmod(37) = \left(5^{12} \bmod(37)\right)^{17} \bmod(37)$$

The secrets to the Diffie–Hellman key generation are the privately chosen random numbers 12 and 17. They must be smaller than, and coprime to, P, that is, not have any common denominators with P.

Eve, an eavesdropper, has listened in to the messages to and fro over the open channel. But even though she is cognizant of both the modulus and the base she is none the wiser. Without knowing either Alices's or Bob's secret number, she cannot compute, divine, or discover the secret key. It would mean computing a discrete logarithm which, for all practical purposes, is impossible for large values of P.[1] (With small P, Eve could launch a brute force attack by simply trying out all possible combinations of secret numbers smaller than P.)

Why does it work? One way to understand the idea behind the key generation scheme is to compare it with the mixing of paint colors. A pitcher filled with yellow color stands on a table in a park for all to see. Alice and Bob come to the park, each carrying an empty can. They fill one-third of

---

[1]   It requires finding the so-called discrete logarithm of 25 to base 5, modulo 50, which is 4. In regular arithmetic it is easy to compute q, for, say $5^q = 625$; thus: $q = \log 625/\log 5 = 4$. But in modular arithmetic $-5^q = 25$ mod(50) – this is impossible when the base (here 50) is very large.

their cans with the yellow color and go home. There they fill their cans up to two-thirds with their secret colors and mix thoroughly. Alice adds red, Bob adds green. They return to the park and exchange cans; Alice hands Bob her can, filled two-thirds with an orange mixture; Bob hands Alice his

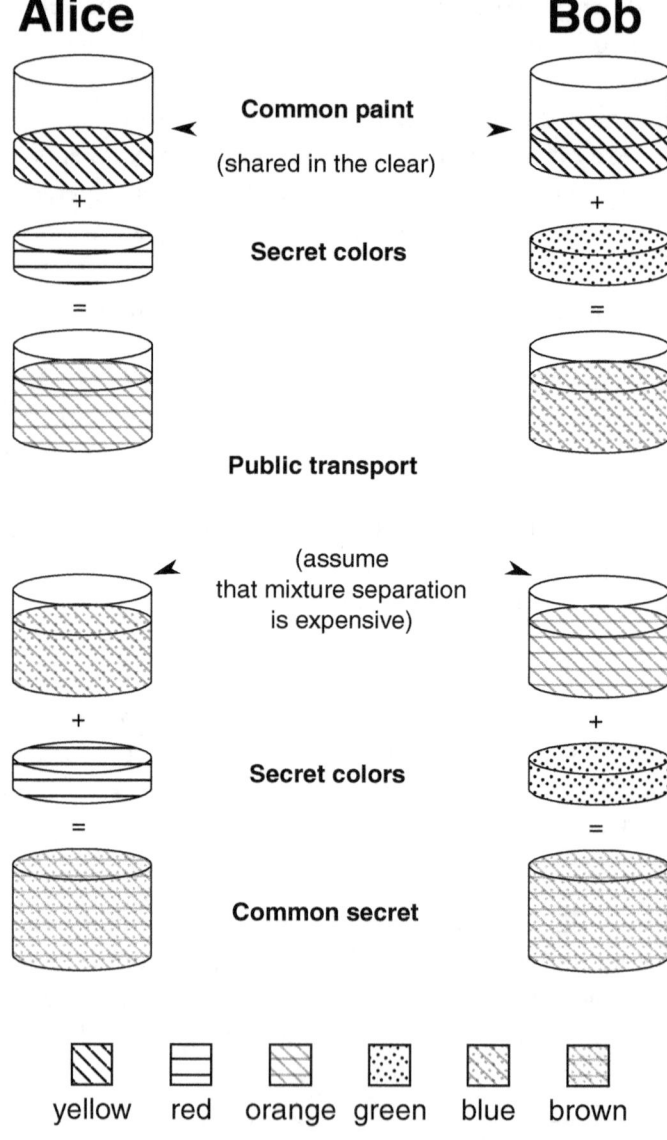

FIGURE 12.1 Diffie–Hellman key exchange (From https://commons.wikimedia. org/wiki/File:Diffie-Hellman_Key_Exchange.svg).

can, filled two-thirds with a bluish mixture. Eve observes all this but cannot make out what color, other than yellow, was admixed.

Alice and Bob go home again, and fill the cans that they just received from each other to the brim, once again with their own secret colors. Alice adds red to the greenish color, mixes it and observes how the content of the can turns brownish. Bob, for his part, adds green to the orange mixture, stirs it and observes how the content of his can turns brownish. Both know that the other person is looking at the very same color. Brown is the secret key. Eve who knows that the base color is yellow and who observed the orange and the greenish mixtures in the park is none the wiser.

Diffie and Hellman's scheme depends crucially on choosing the random numbers for the private keys. For their work, the two scientists were awarded the Turing Prize in 2015, the most prestigious recognition in computer science.

Another secret to encryption is that it is easy to multiply numbers but more difficult to divide them. Multiplying 21 by 17 can be done by most people in their heads. Dividing 357 by 21 is slightly more difficult. (Divisions usually involve guesses and then adjustments whenever there is over- or under-flow.) The situation is comparable to the assignment of telephone numbers. To find the number of, say, John E. Smith is easy for everyone who has a telephone listing and knows the alphabet. But to see to whom the number (212) 345 6789 belongs is impossible, except by reading through the entire listing.

Let's now move to numbers hundreds of digits long. To multiply two numbers that are each about 250 digits can be done by hand. Thought it would take several days to compute the 500-digit product, the important point is that it can be done. Factoring a 500-digit number into its two components is a different matter entirely. It simply cannot be accomplished within reasonable time, not even by superfast supercomputers. Factoring a 75-digit number takes about 100 hours. To factor a 500-digit number would take thousands of years. This is the secret behind the RSA algorithm which we shall describe further.

While the Diffie–Hellman scheme allowed two parties to generate a key that is known only to them – the key can serve as a common password, allow to open a common bank safe, permit access to a database – we now want to see how to encrypt a message. Say Bob wants to transmit his social security number to Alice, without eavesdroppers being able to make it out.

A scheme to do just that was created by three cryptographers just one year after Diffie and Hellman's paper had come out. Ron Rivest from MIT, Adi Shamir from the Weizmann Institute in Israel, and Leonard Adleman from the University of Southern California developed the RSA algorithm, named after the three inventors' initials.

They were the first to publish the scheme but not the first to have invented it. Once again, the British GCHQ had scooped them. Clifford Cocks, who later became GCHQ's chief mathematician had developed the scheme decades earlier, based on findings of his GCHQ colleague James Ellis. Unfortunately, the British powers decided – as always – that their work had to remain classified, though it was shared with the American National Security Agency (NSA). Neither the GCHQ nor the NSA knew what to do with it and never used it, but as government spooks, Ellis and Cocks were not allowed to divulge anything. Their path-breaking roles were only revealed in 1997 which is why their names are practically unknown while the acronym RSA constantly reminds cryptography experts of the later inventors. When Rivest, Shamir and Adleman were awarded the Turing Prize in 2002 the GCHQ mathematicians' names were not on the roster though their priority had at that point in time already been known for five years.

RSA encryption begins with the multiplication of two very large prime numbers, P and Q, chosen randomly, which results in a number N that is generally as long as P and Q combined. P and Q must be prime in order to ensure that N could not – not even theoretically – be factored into more than two factors. (The product of two primes, N in our example, is called a semi-prime.) We compute the number $\varphi = (P - 1)(Q - 1)$. Another number is chosen, E, that must be smaller than, and have no common factors with, φ. Now comes the only difficulty: compute D, such that $D \times E \bmod(\varphi) = 1$. There are algorithms to do that, dating back to Leonhard Euler in the eighteenth century, though they weren't called algorithms at that time.

Alice creates both a public key (E,N) that she may announce publicly for everyone to see and hear, and a private key (D,N) that she keeps strictly secret. (E will denote encrypt, D decrypt.) To illustrate the procedure, we shall use only small numbers in what follows, though for real-life encryption the numbers must be huge. Again, you may skip the following paragraphs.

๛

Alice picks $P = 3$ and $Q = 11$ (hence $N = 33$ and $\varphi = 20$); she chooses $E = 7$. Running Euler's algorithm she obtains 3 for the value of D, since $3 \times 7 \bmod(20) = 1$. So, her public key is (7,33), her private key (3,33).

Bob wants to send the last digit of his social security number (it is a 2) in encrypted form to Alice. He asks Alice for her public key (or looks it up in a publicly available listing). Then he encrypts '2' which is his message to Alice:

$$\text{Encrypted}(2) = 2^7 \bmod(33) = 29$$

The message '29' may be sent over an unsecured telephone line, because, if N is very large, eavesdroppers cannot reckon back from 29 to 2, even if they have knowledge of the public key. Alice, for her part, now uses her private key (3,33) in order to decrypt '29':

$$\text{Decrypted}(29) = 29^3 \bmod(33) = 2$$

Why does this work? It is the shrewd creation of E's counterpart D that allows decryption of the encrypted message. Once Alice has created her public key, (E,N), she used Euler's algorithm to compute D, such that $D \times E \bmod(\varphi) = 1$. Given $E = 3$, the clever creation of $D = 7$ allows decryption of the encrypted message:

$$29^3 \bmod(33) = \left(2^7 \bmod(33)\right)^3 \bmod(33) = 2^{3 \cdot 7} \bmod(33) =$$

$$2^{1 \bmod(\varphi)} \bmod(33) = 2^{1+k\varphi} \bmod(33) = 2 \times \left(2^\varphi\right)^k \bmod(33) = 2 \bmod(33)$$

The last equation follows from Euler's theorem and Fermat's little theorem.

❧

The two schemes that we just discussed hinge crucially on the choice of numbers that are large, prime, and random. Can we be sure that they provide sufficient security? Is it conceivable that the schemes can be broken, either by factoring huge semi-primes into their two prime components or by computing discrete logarithms?

Building on their path-breaking invention of RSA encryption, Rivest, Shamir, and Adleman founded a company in 1982 that they named, unsurprisingly, *RSA Security*. As the internet and online trading became omnipresent, the firm prospered and is today one of the world's largest cybersecurity organizations. Of course, its success hinges crucially on its encryption algorithm and on the confidence that customers have in its resilience. That, in turn, hinges crucially on RSA's claim that, though possible theoretically, it is impossible practically to find the prime components of a sufficiently large semi-prime.

To encourage research into encryption tools using prime numbers, RSA Security ran a series of contests, open to the public. The firm published *RSA numbers*, ranging in length from 100 decimal digits to 617 (330–2,048 bits), and offered cash prizes to anyone who would find the two prime factors. The contests ran for 16 years, from 1991 until 2007.

The first RSA number, consisting 100 decimal digits, was factored into two 50-digit numbers just one month after the challenge was announced. It had taken the Dutch professor of cryptography Arjen K. Lenstra several days on a mini-supercomputer to find the factors.

1522605027922533360535618378132637429718068114961 3806886579084945801229632589528976540003506920061397 = 37975227936943673922808872755445627854565536638199 × 40094690950920881030683735292761468389214899724061

So, the successful factoring of the very first RSA number, corresponding to 330 bits, proved that encryption based on 56-bit keys, as was recommended by the NSA at the time, was not secure. With time, longer and longer RSA numbers were factored and in 2007 the firm ended the challenge: "Now that the industry has a considerably more advanced understanding of the cryptanalytic strength of common symmetric-key and public-key algorithms, these challenges are no longer active."

The largest semi-prime to be factored before the challenge ended was 193 decimal digits long (640 bits) and carried a cash prize of $20,000 and was factored after five months of computation by staff members of the German Federal Office for Information Security. The efforts did not end then, however, and even without cash prizes, teams of computer scientist and cryptographers continue to factor semi-primes. As of this writing, the record holder is a 250-digit (829 bits) semi-prime that was factored by a team of six computer scientists from France and the United States in February 2020. According to a website that reported the feat, it took 2,700 years of computer runtime to carry out the computation, which was done on tens of thousands of machines around the world over the course of a few months.

ॐ

We saw that in order to attack RSA, a semi-prime must be split into its two components. If the semi-prime is large, this is practically impossible. Similarly, recall that an attack on Diffie–Hellman key generation scheme requires finding the integer k, given a, b, and c, such that $b^k = a \bmod(c)$. With large prime numbers c, this discrete logarithm problem is computationally just as intractable as splitting a semi-prime into its two prime components.

Not all encryption schemes require that the secret keys be prime numbers, but they do require that they be random. One example of an encryption scheme that uses random numbers that need not be prime is the Data Encryption Standard (DES), developed in the 1970s by a team of IBM cryptographers.

In 1973, the *Federal Register*, the official gazette of the US government, solicited proposals for cryptographic algorithms and the IBM team – with substantial input from the Bureau of Standards and the NSA – developed the DES, a sophisticated algorithm that was endorsed by the US government in 1977 and soon gained recognition as an international standard. It relied on a single randomly chosen key, both to encrypt and to decrypt information; hence it is called a symmetric algorithm. (In contrast, the RSA scheme and the Diffie–Hellman key generation use different keys to encrypt and to decrypt; they are called asymmetric schemes.)

But criticism quickly arose. Above all, as the RSA challenge had shown, the proposed key length of 56 bits was far too short. It was adopted at the NSA insistence, against IBM wishes. If anything, the involvement of the NSA was ambivalent and led to much suspicion. On the one hand, IBM and the Bureau of Standards wanted a secure system. On the other hand, the NSA, whose mission is to defend the United States against threats from enemies and terrorists, did not like to be hampered in its ability to eavesdrop by overly effective encryption methods.

The NSA representative argued that 56-bit keys were sufficiently safe because at that very time no hardware and algorithms existed that would be able to crack them, while they presumably knew that if the need arose, they would be able to assemble sufficient resources to decipher messages encrypted with 56-bit keys by brute force. To underscore the government's seriousness, the export of encryption algorithms that used keys longer than 56 bits was declared illegal. All that, of course, ran counter to the desires not of enemies of the state or terrorists but ordinary citizens who did not want to grant government agencies a backdoor to their private messages and data.

To further emphasize the claim that 56-bit keys were quite unsecure, RSA Security issued another series of contests between 1997 and 1999, the DES challenge. While the RSA challenge was meant to encourage research into the difficulty of factoring large numbers, the DES challenge was meant to demonstrate that DES, with its 56-bit key, was vulnerable and could be cracked relatively easily.

RSA Security encouraged hackers, geeks, and researchers to decode messages, encrypted with the current DES. The first message was decoded within 39 days, the second, a few months later, within 56 hours. Though this conclusively proved that the encryption methods then endorsed were not secure, the authorities played down the significance. In December 1998, the then-director of the FBI declared:

> That is not going to make a difference in a kidnapping case. It is not going to make a difference in a national security case. We don't have the technology or the brute force capability to get to this information.

But two months later, in January 1999, when the third DES challenge was broken in 22 hours, the powers to be pricked up their ears. To decrypt a message in less than a day was too much to simply hand-wave away. The National Institute of Standards and Technology, successor to the Bureau of Standards, issued a call for a new standard, and in May 2002, after review by the NSA, the *Advanced Encryption Standard* (AES) went into effect, based on key sizes of 128, 192, and 256 bits. To encrypt information classified as *Secret* any of the three keys could be used, for documents classified as *Top Secret*, only the two stronger keys were deemed acceptable. Encryption methods to protect information of *national security* required further review.

The emphasis in this chapter was on secret numbers, be they prime or not, that must be chosen randomly. For RSA encryption, two prime numbers, each approximately 1,024 bits (300 decimal digits) long, are randomly chosen and multiplied, to give a 2,048-bit (600 decimal digits) semi-prime. As of this writing, this is still considered safe but, with the advent of new technologies, the keys may have to be significantly lengthened. Keys for Diffie–Hellman key generation are currently between 1,024 and 2048 bits long and their security is as strong as the RSA with comparable keys. AES keys are between 128 and 256 bits (between about 39 and 77 decimal digits) long. To crack AES by checking each of the possible key values would take billions of years.

To summarize, AES, being a symmetric key algorithm, uses large random (not necessarily prime) numbers for encrypting data efficiently once a shared secret key is established. In contrast, RSA and Diffie–Hellman, as asymmetric key algorithms, use very large prime numbers, also randomly chosen, for securely exchanging keys, public key encryption, and authentication purposes.

❧

But there's an elephant in the room: we described some methods to choose random numbers in the previous chapters. But how does one randomly choose a *prime* number, several hundred decimal digits, or several thousand binary digits long, if one wants to use the RSA or Diffie–Hellman algorithms?

The prime number theorem, proved in the late nineteenth century, independently by the mathematicians Jacques Hadamard and Charles Jean de la Vallée Poussin, gives us an indication. It says that the number of prime numbers smaller than N is approximately $N/\ln(N)$. (ln is the natural logarithm; in what follows, all numbers are approximate.)

Let us say that we want to pick a random prime number that is 300 decimal digits (about 1,000 bits) long. First of all, how many numbers are there to choose from? The answer is

$$10^{300} - 10^{299} = 9 \times 10^{299}$$

Now let us see, based on the prime number theorem, how many of the numbers are prime:

$$10^{300} / \ln\left(10^{300}\right) - 10^{299} / \ln\left(10^{299}\right) = 10^{300} / 690.7755 - 10^{299} / 688.4729$$
$$= 1.3018 \times 10^{297}$$

So, the fraction of 300-digit numbers that are prime is approximately

$$1.3018 \times 10^{297} / 9 \times 10^{299} = 0.1446\%$$

or about 1 in 700.

The following algorithm produces a random prime number: a random 300-digit number is generated by any of the methods that have been discussed in previous chapters, or that will still be discussed. The crucial next step is to determine whether the generated number is, in fact, prime. This can be done using probabilistic primality tests such as Miller–Rabin primality test that we will discuss in more detail in Chapter 15. These tests can quickly identify non-prime numbers and dismiss them, which is what happens in about 98.554% of the cases. When the number is not dismissed, it becomes a candidate but since Miller–Rabin test occasionally misidentifies a composite number as prime, the candidate must be re-tested several times to reduce the probability of error. If the candidate turns out to be composite, another 300-digit random number is generated and the process is repeated until a prime number is found.

The entire process is fast: each Miller–Rabin test requires about two milliseconds. On average, about 700 numbers are tested. Once a candidate is identified, it is re-tested 10 or 20 times in order to reduce the probability of mis-identification to a minimum. Altogether, the procedure takes about 1.5 seconds.

Cryptography, in online trading, exchange of private messages, maintenance of databases, is one of the best-known applications of random numbers. In the next chapter, we discuss another cryptographic application that requires random numbers: zero-knowledge proofs allow one party to prove to another party that a statement is true without revealing any additional information. This ensures privacy and security in digital interactions, such as secure authentication, and confidential data sharing, by allowing the receiving party to verify information without requiring access to the underlying data.

CHAPTER 13

Compute π by Throwing Darts

The Monte Carlo Method

In a previous chapter, I described the methods that the so-called dynamical systems collective employed in the casinos of Las Vegas in order to beat randomness at the game of roulette. More elegant than the gambling dens in Nevada is the celebrated casino in Monte Carlo, on the shores of the Mediterranean. So, let us now talk about what became known as the Monte Carlo method. Actually, though the method makes abundant use of randomness, it has little to do with the casino in the Principality of Monaco itself. The only connection is that one of the early proponents of the method suggested the name because an uncle of his would borrow money from relatives because he "just had to go to Monte Carlo."

Essentially, the Monte Carlo method is a protocol for testing the characteristics of a complex system by using random sampling. Performing a large number of random simulations, and then aggregating the results allows one to create a statistical representation of the system. The method provides accurate estimates even when the system is too complex to be described perfectly by its underlying principles or when those principles are unknown. For the Monte Carlo method to be effective, two conditions must be met: first, a large number of simulations must be conducted;

DOI: 10.1201/9781003641520-18

second, these observations must be based on the use of random numbers to ensure unbiased sampling.

᷍

In the nineteenth century, two different mathematical methods were used to analyze physical phenomena.

> Problems involving only a few particles were studied in classical mechanics, through the study of systems of ordinary differential equations. For the description of systems with very many particles, an entirely different technique was used, namely, the method of statistical mechanics. In this latter approach, one does not concentrate on the individual particles but studies the properties of sets of particles.
>
> (Metropolis, 1987)

Statistical mechanics, a well-developed branch of physics, founded in the nineteenth century by Ludwig Boltzmann and others is what led to what would become known in the late 1940s as the Monte Carlo method. Actually, the Italian physicist and later Nobel Prize winner Enrico Fermi was the first to use the method already in the 1930s, before it actually had a name. This was much before the times of electronic computers and all that Fermi had at his disposal was a mechanical adding machine which he used during sleepless nights to study problems in neutron diffusion. These problems could not be solved with traditional mathematical techniques, so he had recourse, without telling anyone, to the computation intensive methods of statistical mechanics. The results he obtained accorded remarkably well with experimental results and he "took great delight in astonishing his Roman colleagues with his remarkably accurate, 'too-good-to-believe' predictions of experimental results. After indulging himself, he revealed that his 'guesses' were really derived from the statistical sampling techniques… ." Fermi had independently developed the Monte Carlo method nearly 15 years before its official inauguration.

So what is the Monte Carlo method, a.k.a. the Monte Carlo simulation?

Let us say, we want to find the numerical value of the number π. We know that a circle has a surface of πr^2, so a circle with a radius of 1 cm has a surface of π cm^2. We inscribe the circle into a square of side length 2 cm; the square has an area of 4 cm^2.

Let us look at the top right quadrant of this square; it has an area of 1 cm^2 and the quarter circle inscribed in this quadrant has a surface of $\pi/4$ cm^2. Now we throw darts randomly at the quadrant. The proportion of darts that land inside the quarter circle to all the darts that were landed inside the top right quadrant is equal to $\pi/4$. If we randomly throw a thousand darts,

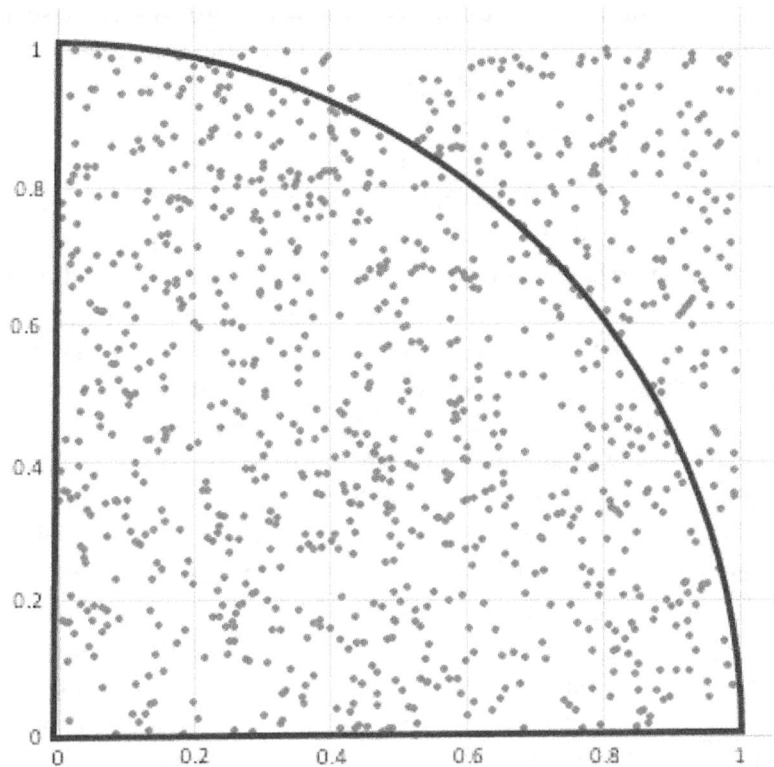

FIGURE 13.1   Simulating the number π, ©gsz.

usually 785, plus or minus a few, will land inside the quarter circle. What does that tell us?

We can recreate the dart throws with random numbers. Choose two random numbers between zero and one. These are the coordinates of a point in the quadrant of the *xy*-plane. By the mathematical definition of a circle, if the sum of the squares of these two numbers is less than 1.0, the point lies inside the circle, if it is greater than 1.0 it lies inside the square but outside the circle.[1] For example, the point indicated by the pair 0.53 and 0.27 lies inside ($0.53^2 + 0.27^2 < 1$), while the point indicated by the pair 0.85 and 0.76 lies outside ($0.85^2 + 0.76^2 > 1$). Repeat this exercise many, many times, compute the proportion, and you will get closer and closer to 78.53…%, which is equal to π/4.

---

[1]   A circle of radius *r* is defined as all points with coordinates *a* and *b*, such that $a^2 + b^2 = r^2$. The inside of a circle of radius *r* is defined as all points, such that $a^2 + b^2 < r^2$.

Finding the numerical value of π is a simple Monte Carlo exercise in the *xy*-plane, that is, in two-dimensional space. (You, the reader can easily do it yourself with a simple spreadsheet.) But mathematicians like to wander around in high-dimensional spaces. Not only they, but engineers too, and scientists, often deal with spaces that have high dimensions. For example, a space ship has a position in regular *xyz*-space, but also exhibits velocities and accelerations in all three directions. Thus, its state at any given moment can be represented by a point in nine-dimensional space. If we define limits on the ranges in each dimension – for example, at speeds above p in direction x acceleration in direction y must be less than $\sqrt{z}$, and so on – we have defined a high-dimensional body. To compute the volume of this body with traditional mathematical methods is difficult. Much simpler it is, to draw many, many sets of nine random numbers, check how many of the points defined by these sets fall within that body, and compute their proportion to all points. In this manner, volumes of 9-, 10-, 20-dimensional bodies can in principle be estimated. (But see further for a new problem that arises in this case: the curse of dimensionality.)

Of course, many, many, many random numbers are needed for a useful implementation of this method. To wit, my Monte Carlo simulation for π, with 10,000 pairs of random numbers performed on a Microsoft Excel spreadsheet produced an estimate of 3.1232..., an error of about half of 1%. While that would be acceptable for general usage, for mission-critical calculations the error is far too large. My simulation with one million pairs of random numbers on a Microsoft Excel sheet resulted in an estimate of 3.144824..., an error of only 0.1%. The more pairs of random number one uses, the more exact the result will be.

Since Fermi never published anything on statistical mechanics, nobody knows how he came up with sufficiently many random numbers in the 1930s, and his trick lay dormant for about a decade. By then, World War II was raging and the Manhattan Project, the American effort to develop the atomic bomb, was in high gear.

One of the tasks the Manhattan Project scientists at the Los Alamos National Laboratory in New Mexico had to tackle was to analyze the behavior of nuclear reactors. They sought to understand the motion and interaction of neutrons within materials...where they were, what direction they were going, how quickly they moved. To solve such problems by the classical approach, with differential equations, was not possible and Fermi's pre-war technique was revived. Leading the efforts to apply the methods of statistical mechanics to the problem were the physicist Nicholas Metropolis, originally from Greece, and the mathematician Stanislaw Ulam, originally from Poland. John von Neumann, originally from Hungary, considered one of the foremost mathematicians of the twentieth century, was also instrumental in the effort.

With the war over, statistical mechanics was in danger of falling into oblivion again due to the difficulties in generating sufficient amounts of random numbers. But the late 1940s were the dawn of the era of electronic computers and they are what permitted mathematicians and physicists to make the Monte Carlo method an indispensable tool that it is today. When Ulam encountered ENIAC at the University of Pennsylvania, the first programmable, electronic digital computer (weighing more than 27 tons and covering 167 m^2 of floor space) he was duly impressed by its speed and versatility. "Stan's extensive mathematical background made him aware that statistical sampling techniques had fallen into desuetude because of the length and tediousness of the calculations," Metropolis recounted.

> But with this miraculous development of the ENIAC – along with the applications Stan must have been pondering – it occurred to him those statistical techniques should be resuscitated, and he discussed this idea with von Neumann. Thus was triggered the spark that led to the Monte Carlo method.

Key to the application of the Monte Carlo method is the ability to generate vast amounts of random numbers. If in 1958, an engineer marveled that the newfangled electronic thingies of his day could generate 200 random numbers per second, suffice it to say for now that as of this writing, the fastest computers were able to generate 250 terabits (250,000,000,000,000 zeros and ones) per second which corresponds to between 40 and 75 trillion decimals per second. (In the appendix to this chapter, I show how random sequences of bits can be converted into random sequences of decimals.)

We saw that the Monte Carlo method allows us to approximate the numerical value of π, estimate the volumes of convoluted bodies, investigate the behavior of neutrons. What else can we do with the Monte Carlo technique?

When engineers work on a project, or when scientists plan an experiment, or when economists evaluate a policy, they have a model on paper that describes their theory, complete with equations, statistical information on the variables, possible values of the parameters, conceivable impact of environmental factors, and more. To evaluate the project with different input data, computations must be performed that are often far too complex to be solved analytically. As was the case with neutron diffusion, the way out is to observe situations virtually, that is, to simulate them on computers.

Model complexity is not the only reason to resort to simulations. Sometimes a model cannot be tested because it would be too cumbersome to actually construct it. Or an actual test – a nuclear explosion, for example – may be out of the question. Or an experiment – testing vaccines on children – may be morally unacceptable. Or the real thing – a space station – may be inaccessible. Or a test may destroy the object. Or errors may be too costly.

In such situations, the projects must be tested *in silico*, before they are actually constructed. Depending on the nature of the project, there are different requirements. Tunnels, aircraft, self-driving cars must be safe under all circumstances and be designed to withstand all adverse events. Hence, worst-case scenarios are considered. A bridge must not only be able to endure the fairly rare occurrence of force-four hurricanes or the once-in-a-century magnitude-seven earthquake but, importantly, must withstand the even rarer worst-case scenario of both events occurring simultaneously. To design the bridge merely for what may happen in an average year would be reckless.

On the other hand, an architect's best guess when planning a housing project for 150 families may be that the families have 1.8 children on average, 2 out of 3 families have a car, and every family has one bicycle. So, she will include a playground for 270 children, a garage for 75 cars, and stands for 150 bicycles. Her design decisions are dictated by these best guesses. To design the project for a worst-case scenario – say five kids, three cars, and four bikes per family – would be unnecessarily expensive and wasteful.

The engineer's and the architect's decisions are simple enough: each uncertain variable is assigned a unique value – either 'worst case' or 'best guess' – and the design requirements follow from there. But some projects may require a more subtle approach.

The manager of an automobile assembly plant wants the operation to be efficient: maximize output, minimize waste, and avoid costly work stoppages. There are parameters under his control and many variables which are not. Let's say that he can set the speed of machine A either to six units or to nine units per hour. Then, for each of the many variables that he cannot control there is a whole array of possibilities, each of which comes with a certain probability. There are usually 15 workers on the floor but between 1 and 3 fall sick twice a month; the screws on machine B get loose once a week and it takes between 20 minutes and 1 hour to tighten them; when machine C overheats, which happens about every 3 months, it has to cool down for up to half a day; when machine D breaks down, which occurs very rarely, the workflow is diverted to machine E which is slow but reliable. And so on, and so on. There are dozens of variables, each with its own probability distribution, and thousands of different combinations of incidents.

The manager must decide whether to set the machine A to the high or the low throughput. To compute the expected cost, allowing for the impact and interactions of all uncontrollable variables, would be prohibitively cumbersome. And this where Monte Carlo simulations come in.

Sets of four three-digit random numbers indicate what scenario is being simulated. If the first number is less than 500, 15 workers show up; if it is between 501 and 950, 15 workers show up; and if it is between 951 and 999, just 13 come to work. If the second random number is greater than 860 but less than 881, a screw got loose in machine B and it takes 20 minutes to tighten it; if the number is between 882 and 982, it takes 45 minutes; if it is between 982 and 999, 2 hours. The third number indicates whether machine C overheats, the fourth whether machine D breaks down.

In this manner, many, many scenarios are created virtually. Each set of four random numbers allows the manager to simulate a situation that can actually occur, together with its probability. A well-designed Monte Carlo simulation will produce specific scenarios approximately as frequently as they would appear in real life...some more often, some less often, some more serious, some less so. By running through very many possible scenarios, the Monte Carlo technique allows to simulate what would occur naturally, without ignoring the low probabilities of catastrophic events.

Using truly random numbers is crucial in simulations. Analyzing the structural integrity of buildings, for example, under various loads such as wind, earthquakes, and weight by using non-random, predictable numbers would test the structure only against expected, non-variable loads. This could lead to underestimation of vulnerabilities, for example, when several stress events coincide. By running simulations with random numbers, one creates a more robust design that can account for a wide range of potential scenarios.

Traffic flow simulations that rely on random numbers are important in the design of efficient road networks and traffic light timings. If simulations rely on fixed traffic volumes at specific times, the system may appear optimal but fail during unexpected surges. Randomized traffic patterns help simulate real-world conditions, such as accidents or public events, leading to more adaptable and efficient traffic systems. Comparing traffic patterns under predictable versus random incident scenarios demonstrates how systems designed with variability in mind can better handle real-world challenges.

For simulation algorithms to work properly and in an unbiased manner, it is crucial that every number has the appropriate chance of being selected. Usually, this is achieved when the random numbers are uniformly distributed and the probability of any number (zero to nine) or bit (zero or one) being selected is equal. But some simulations might require random numbers that follow different distributions. For instance, normal distributions (Gaussians or bell curves) are commonly used in simulations involving natural phenomena or financial simulations where values are clustered around a mean and taper off at the extremes. In simulations of queueing systems (like customer service centers, network routers, etc.), service times might be modeled using exponential or Poisson distributions to reflect the random nature of service durations and arrival times. The famous Black–Scholes model of options pricing assumes that the logarithmic returns of a stock price follow a normal distribution. Financial risk models might use distributions like the t-distribution to simulate extreme markets. In the simulation of insurance scenarios, the distributions of claims and losses often have heavy tails to account for rare but severe events (e.g., natural disasters, large-scale accidents). Simulations may use the Pareto distribution for such purposes.

I mentioned the usefulness of the Monte Carlo method to compute the surfaces of irregular shapes and the volumes of complex bodies. But for high-dimensional bodies there's a problem; it is known as 'the curse of dimensionality.' The problem is that randomly distributed points become very sparse in high-dimensional spaces.

Consider the unit cube in three-dimensional space. It has side-lengths of 1 cm in all three directions, and the volume is 1 cm³. Now consider a sub-cube with side-length 0.5 cm in all directions. The volume of this sub-cube is one-eights of the volume of the unit cube. A simulation with one million triples of random numbers (triples, in order to define points in three-dimensional space) would place about 125,000 points within the sub-cube. Hence, we would estimate the volume of the sub-cube correctly as $^{125,000}/_{1,000,000} = ^1/_8$ of 1 cm^3.

Now, let us elevate this to 20 dimensions. A sub-cube with side-length 0.5 in all 20 directions has a volume of $^1/_2{}^{20}$ of the unit cube, that is, about $^1/_{1,000,000}$. Hence a Monte Carlo simulation with one million random points, each point defined by 20 random coordinates (altogether 20 million random numbers) will only have, on average, one point in the sub-cube. But it could contain two, five, zero points … . Depending on whether the random points happen to be clustered or sparse at that very location, the resulting simulation could be completely incorrect. This is the curse of dimensionality.

Well, if truly random numbers are inadequate for moderately high-dimensional situations, the so-called *quasi-random numbers* come at least partially to the rescue. These numbers are characterized by the fact that they fill every sub-cube according to the proportion of its volume, forming no clusters and leaving no empty regions. They lie somewhere in between ordered numbers and random numbers.

How can quasi-random numbers be obtained? Not by coin tosses or dice throws because our otherwise beloved random numbers could throw off high-dimensional Monte Carlo simulations if sub-spaces are visited too sparsely or too abundantly. We don't want the space to be interspersed with holes like Emmental cheese.[2] What we require are numbers that are random, but evenly spread out. Now, if that sounds like a jarring contradiction, it is…which is why a compromise must be found.

There exist algorithms that generate quasi-random numbers. They strive to generate numbers such that if they are interpreted as coordinates of points, these points fill all sub-spaces with a proportional number of points. If a three-dimensional space encompasses, say, a volume of 10 cm^3, then every 1 cm^3 sub-space should contain approximately one-tenth of all points.

To illustrate, I describe one well-known procedure that generates quasi-random points n_i $(i = 0,…N)$ in the two-dimensional unit square. It is named

---

[2] Please do not call the holey cheese 'Swiss cheese.' There are dozens of different Swiss cheeses, and only the 'Emmentaler' has holes.

after the Dutch mathematician Johannes van der Corput (1890–1975) and is so complicated and intransparent as to border on numerology. A well-known mathematics commentator remarked that this "bizarre mashup of operations on numbers and operations on numerals" is "bit-twiddling" that is "too weird for words." First, mark n_i evenly spaced points in the interval N. The n_i/N will be the x-values (if $N = 8$, 6 becomes $^6/_8$). To obtain the corresponding y-value, convert n_i to base 2 (e.g., 6 becomes 110), invert the sequence (110 becomes 011), re-convert the result to a decimal (011 becomes 3), divide by N (3 maps to $^3/_8$). So, the quasi-random point in the unit square will have coordinates $^6/_8$ and $^3/_8$.

Of course, the van der Corput points are nowhere near random. For one, the x-coordinates are evenly spaced between zero and one ($^0/_8$, $^1/_8$, $^2/_8$, ...,$^7/_8$). And the y-coordinates, though seemingly random, are completely deterministic. Nevertheless, quasi-random numbers like the ones just described do give good results for Monte Carlo procedures in moderately high dimensions. The reason is that the points fill the unit square without obvious holes.

While true random numbers fulfill the sacred trinity of randomness – they must be *u*npredictable, *u*niformly distributed, and *u*ncorrelated (U³) – quasi-random numbers do not even pretend to satisfy the third requirement; they are not at all independent of each other. In fact, in order to fill spaces left open, they avoid clusters by 'remembering' where the previous points lie.

For really high-dimensional spaces, even the van der Corput technique does not suffice. The points obtained by random sampling and even by quasi-random sampling are too sparse to provide useful results. To estimate the volume of very high-dimensional bodies, a variation of the Monte Carlo technique was developed by Nicholas Metropolis and four colleagues in the 1950s.[3] Their algorithm starts with a point that lies inside the body and then steadily inspects further points within a certain neighborhood of the previous point that are chosen randomly according to a certain probability distribution. Simulating a random walk in high-dimensional space, the algorithm allows an estimation of the body's volume without falling prey to the curse of dimensionality, by computing the proportion of inspected points that lie inside the body.

---

[3] The co-authors were two married couples: Edward and Augusta Teller and Arianna and Marshall Rosenbluth.

The possible applications of Monte Carlo simulations are legion: telephone traffic, urban planning, spread of virus, stock market performance, behavior of atoms and molecules, appraisal of investments, valuation of corporations, computer and video games. They also help to prepare airline pilots for surprising situations, coach surgeons to deal with unexpected eventualities, train autonomous vehicles for unforeseen situations, all of which, at that point in time, exist only on paper or, rather, *in silico*.

APPENDIX: CONVERTING RANDOM BITS INTO RANDOM DECIMALS

There are two methods to convert bits into decimals:

(1) Method 1 partitions the entire bit-sequence into groups of four:

1001010011010110 becomes 1001 0100 1101 0110,

and the groups are converted into decimals:

$$1001 \Rightarrow 9, 0100 \Rightarrow 4, 1101 \Rightarrow 13, 0110 \Rightarrow 6$$

Decimal numbers greater than 9 and above (like $1101 = 13$) are discarded, and we obtain the random sequence of decimals 9, 4, 6.

A drawback of Method 1 is that only about 62.5% ($= {}^{10}/_{16}$) of the groups result in decimals between 0 and 9; about 37.5% of the groups represent numbers 10 to 15 and are discarded.

So, 250 terabits of random zeroes and ones would result in about 40 trillion random digitals:

$$(250,000,000,000,000 / 4) \times 0.625 \sim 40,000,000,000,000$$

(2) Method 2 converts the entire sequence into a decimal numbers:

$$1001010011010110 \Rightarrow 37,654$$

and we obtain the random decimal sequence 3, 7, 6, 5, 4.

This method produces more decimals since no bits are discarded. Since a decimal digit is represented by approximately 3.32... bits ($=\log_2 10$), 250 terabits correspond to about 75 trillion decimal digits:

$$(250,000,000,000,000 / 3.32...) \sim 75,000,000,000,000$$

But it also has a drawback: converting a very long binary number to a decimal number might involve iterative steps or the use of floating-point arithmetic, where precision is limited; truncation and small rounding errors at each step can accumulate. Furthermore, if the binary number exceeds the representational capacity of standard data types, the conversion might result in overflow. Hence, there is no guarantee that the individual decimal digits are uniformly distributed.

To summarize: though Method 2 generates more decimals, Method 1 results in the decimals being uniformly distributed.

CHAPTER 14

"You can fool all the people half the time ..."

Zero-Knowledge Proofs

Having explored the use of random numbers in gaming and gambling, polling, sampling, making money, encryption, simulation and even choosing office holders, we now describe a rather esoteric application, namely how to convince someone, using random bits, that one knows a secret, without divulging the secret.

In the early sixteenth century, it was received wisdom that no formula existed to solve cubic equations, $x^3 + ax^2 + bx + c = 0$. But then the Italian mathematician Niccolò Tartaglia (1500–1557) unexpectedly claimed that he had discovered the formula that solves cubics but would keep it a closely held secret. He was the sole person in the world who knew how to solve cubic equations, and he intended to keep it that way. A compatriot, Gerolamo Cardano (1501–1576), was very intrigued. As he was working on a book about arithmetic, he was about to claim that cubics could not be solved.

How could Tartaglia prove to Cardano that he knew the secret formula, but without divulging it? Easy: he told Cardano to send him a few dozen cubics, and by return mail sent him the solutions ... albeit without letting on how he arrived at them. Baffled, Cardano was easily able to verify the correctness of the answers by plugging them into the cubics but was left

DOI: 10.1201/9781003641520-19

frustrated at not knowing how Tartaglia had done it. There's more to the story – accusations of heresy, plagiarism, double-crossing, broken oaths – but I won't let myself be sidetracked here.

The story illustrates an aspect of what would become known as *zero-knowledge proofs* (ZKP): Tartaglia proved to Cardano that he knew the formula that solves cubics. And Cardano realized that Tartaglia was telling the truth, albeit without learning anything about the formula itself. The protocol required interaction between the prover (Tartaglia) and the verifier (Cardano) in the form of challenges sent by the latter to the former. Any eavesdropper to the interaction between Tartaglia and Cardano would also realize that Tartaglia knew the secret formula, also without gaining any further information.

Another example of a proof that does not divulge secret information is a secure multiparty computation (MPC) protocol that distributes a computation across multiple parties, without any individual party seeing the other parties' data. It allows not only to prove that something is true but also to compute the answer to a certain question without revealing any information about the inputs.

Let's say three office workers, Alice, Bob, and Carol, want to find out what their average salary is without revealing their individual salaries. Of course, a trusted third party, say Xavier, could collect the three salaries, compute the average, and share the result without disclosing anything else. However, this approach means that one must trust Xavier which may not always be desirable. Maybe random numbers can be utilized to bypass Xavier?

An MPC protocol would proceed as follows: Alice chooses a random number, adds it to her salary and sends the result to Bob. He also chooses a random number and adds it and his salary to the sum he received from Alice and transmits the result to Carol. Carol, in turn, adds her salary and *her* random number to the sum and sends the sum – which now consists of the three salaries and the three random numbers – back to Alice. Alice deducts her random number from the sum and sends the result to Bob. Bob deducts his random number and sends the result to Carol. Finally, Carol deducts her random number and announces the result. Divided by three, this is the average salary which is now known to all three, without any of them knowing the other two salaries.

✍

It was at a symposium in Rhode Island in 1985 that ZKPs were born. The Israeli computer scientist Shafi Goldwasser, the Italian Silvio Micali, and the American Charles Rackoff presented "The Knowledge Complexity of Interactive Proof-Systems," a paper that was specifically mentioned in the laudation when the first two authors, Goldwasser and Micali, were awarded the Turing Prize in 2012. The paper introduced *knowledge complexity* as a measure of the knowledge that the prover transfers to the verifier in the course of her proof.

There are many instances in life when one may want to prove a fact without actually divulging it. As we just saw, you may want to keep your salary secret; or you may want to prove to a bank that you have a good credit score, without actually disclosing it; you may want to convince an employer that you do not have a certain health condition, without revealing your complete medical history; you may want to show the bartender that you are of drinking age, without revealing how old you are...

It has become a convention among the cognoscenti that the prover is named Peggy and the verifier Victor. So, to illustrate Goldwasser and Micali's ZKP, let us imagine the following scenario:

Eve lost her smartphone. She is sad because she had installed two home screens, one that showed a Teddy, the other a Mickey. Whenever she got tired of seeing the one, she could type in her personal identification number (PIN) and the screen would switch to the other. Otherwise, the phone randomly switched between Teddy and Mickey.

Peggy found the phone on a bench in Central Park and tells Victor that she managed to hack into the phone and discover the PIN. To prove to Victor that she had really been able to do so, without divulging the number itself, she proposes a game. Sitting opposite Victor with the back of the smartphone facing him, she challenges him to shout out 'Teddy' or 'Mickey' and she would show him the appropriate picture.

To begin the game, Peggy looks at the screen; a Teddy is displayed. Now Victor challenges her and after thinking a second, calls out 'Teddy'. Peggy turns the phone around to show Victor the Teddy. Nice! But what if Victor had called out 'Mickey?' Well, Peggy would have keyed in the PIN to flip the picture, and then shown the Mickey.

But Victor is not convinced. Even if Peggy had not known the PIN, there would have been a 50% chance that she would have been able to show the correct picture anyway. If Mickey had been displayed at the outset, and Victor just happened, by coincidence, to call out 'Mickey', she would have fooled him. Hence, Victor cannot be sure; there is a 50% probability that Peggy showed him the correct picture by sheer luck.

But Victor has a thing up his sleeve. By challenging Peggy several times, he can decrease the probability of being fooled. Each time Peggy shows the correct picture, the probability that she was able to do that purely by coincidence is reduced by one half. (Recall that, left on its own, the phone randomly switches between Teddy and Mickey.) If Victor challenges Peggy n times, there is only a 2^{-n} chance that Peggy, by pure luck, displays the correct picture every time. To paraphrase a famous quote, attributed to Abraham Lincoln, "you can fool all the people half the time and some of the people all the time, but you cannot fool all the people 2^n times."

The crucial point of the scheme is, of course, that the challenges that Victor shouts out are purely random. If they are, and Peggy shows the correct home screen every time, she will have convinced him – up to, or rather down to, a negligible probability – without having divulged any relevant information beyond that.

✑

Among the many applications of ZKPs, the procedure has also been proposed to solve an issue of utmost international importance, namely ascertaining a nation's nuclear capabilities without actually bringing a bomb to explosion. In June 2014, the aerospace engineer Alexander Glaser, the computer scientist Boaz Barak, and the astrophysicist Robert J. Goldston suggested a protocol that would allow this. I will describe a simplified version of their proposal.

Abulafia, a superpower in the North, is known to possess a new kind of weapon. Its scientists have mastered the technique of aggregating the top-secret amount of the very volatile material pluranium (52.5 kg). Buzaglio, an emerging country in the South, claims also to have mastered the technique. How can Buzaglio prove to AVI, the *Agence pour la Vérification Internationale*, that it has in fact produced ten warheads, without divulging the top-secret amount?

Let us say that Buzaglio tells the truth. The verification procedure goes as follows:

Abulafia provides one of its true warheads, the reference warhead REF. Buzaglio provides the ten warheads it has manufactured (WAR_1 to WAR_{10}) and two dummies (DUM_1 and DUM_2) that have the identical mass, but their weight is also kept secret. AVI tells Buzaglio to place one dummy and WAR_1 into one pan of a beam scale, and the other dummy and Abulafia's REF into the other pan. The beam balances, so both warheads weigh the same (70 kg). The exercise is repeated for WAR_2 to WAR_{10} with the dummies,

DUM_1 and DUM_2, chosen at random for each pan. Every time the weights match and AVI has been convinced that Abulafia and Buzaglio's warheads have identical weights. But, crucially, AVI does not know the secret amount of pluranium.

What if Buzaglio lies and has never been able to aggregate more than 40 kg of pluranium before the device would explode. Can it fool AVI? Well, yes, but only with a minuscule probability.

Buzaglio provides the ten fake warheads ($FAKE_1$ to $FAKE_{10}$) that contain only 40 kg of pluranium, and two dummies. But this time the dummies have different weights: DUM_1 weighs 17.5 kg and DUM_2 weighs 30 kg. AVI points to one of the dummies and tells Buzaglio to put it into the beam scale's pan together with REF, and the other dummy together with $FAKE_1$ into the other pan. Lo and behold, they both weigh the same (70 kg). But that happened only by coincidence because it was DUM_1 that was placed together with $FAKE_1$, and DUM_2 with REF. Had AVI pointed to DUM_2 and told Buzaglio to place it together with $FAKE_1$, and DUM_1 together with REF, the beam would have tipped. (The weights would have been 57.5 and 82.5 kg.)

So, the first trial worked, albeit with a 50% probability of having given a wrong result. The exercise is repeated for the other nine warheads. Each time, AVI points randomly to one of the two dummies and tells Buzaglio to place it together with REF, and the other together with $FAKE_n$. And each time, there is a 50% chance that the weights match even though the warhead is fake. But the probability that *all* weights match, is 2^{-10}, that is, less than one-tenth of 1%. Crucially, AVI must choose randomly between the two dummies.

To summarize: on the one hand, random selection of dummies and repetition of tests ensure that Buzaglio cannot consistently deceive AVI; the probability of passing all tests by chance if Buzaglio is lying is negligible. On the other hand, AVI verifies the weights of the warheads but remains ignorant of the exact amount of pluranium used.

The protocol proposed by Glaser, Barak, and Goldston may become useful not only to verify whether a nation possesses nuclear bombs but also to get rid of them. For the purposes of disarmament, the protocol could ascertain that bombs that are to be destroyed are actual nuclear devices and not just fakes.

❧

As if the fact that the truth of a statement can be established without revealing the proof itself were not astonishing enough, there is another surprise in store. In 1991, two Israeli and an Italian computer scientists published a result that would become very influential. Based on prior work, they proved (yes: proved!) first, that any statement whose correctness can be efficiently verified can be converted into a sort of map (like a geographical map with many countries); second, that if the statement is true, that map is colorable with only three colors, such that no neighboring areas have the same colors; and, third, that the fact that this map is *three-colorable* is provable in zero knowledge! Hence, if a proof for a mathematical hypothesis with a verifiable solution exists, the hypothesis's correctness can, in principle, be translated into a map and proved in *zero knowledge*. Hypotheses (e.g., the Riemann hypothesis, the twin prime conjecture, etc.) would become theorems though nobody, except the provers, would know the particulars of the proofs.

ZKPs extend the traditional notion of a proof from a static text to a protocol, which involves a sequence of random numbers and interaction between the prover and verifier. To qualify as a ZKP, the protocol must fulfill three requirements: it must be complete, sound, and zero knowledge. It is complete if an honest prover can convince an honest verifier, up to a negligible probability, that a true statement is true. It is sound if a cheating prover cannot convince an honest verifier of a false statement, except with a negligible probability. And it is zero knowledge if a verifier learns nothing except the fact that a statement is true.Applications of ZKPs are legion: privacy-preserving data mining, online auctions, validation of cryptocurrency transactions, attribution of cyberattacks without revealing classified intelligence, patent disputes while keeping the patent itself secret, proving origin of data without revealing how it was obtained, and more.[1]

Crucially, for ZKPs to function effectively, the shout-outs in the examples given, Victor's 'Teddy' or 'Mickey,' and AVI's 'DUM_1' or 'DUM_2,' must be a random sequence of bits. Paradoxically, by ensuring verification without revealing sensitive information, ZKPs manage to reconcile two seemingly contradictory concepts: while transparency is typically anathema to security, ZKPs achieve transparency by providing information, albeit maintaining security by withholding information.

---

[1] An aside: in the game of poker, the winner must provide opponents only with partial knowledge; he or she must disclose only the lowest hand that beats the opponents, without revealing one's true hand. And someone who is good at bluffing need not reveal the hand at all if he or she convinced all opponents to fold. Losers do not know if the winner really had a higher hand or was only bluffing. They are left with zero knowledge, except for the fact that they lost their stakes.

CHAPTER 15

Deus ex Machina

Randomized Algorithms

Received wisdom is that computer algorithms are fast and that at every moment the next step is strictly determined. However, this is only partly true. Deterministic algorithms may exhibit bad behavior in worst-case scenarios. For example, when the input becomes large, running time increases, sometimes exponentially so. In these cases, it may be advantageous to forego strict determinism and introduce randomness instead. Like the *deus ex machina* in ancient Greek tragedies – to resolve complex problems or dilemmas within the plot, an omnipotent deity was lowered onto the stage with a crane-like device (*mechane*) – random numbers are the surprising outside power that come to the rescue.

Exploring random numbers through, say, cryptography and the Monte Carlo method has helped us see just how powerful they can be in the right context. But there is more to random numbers than that; they can also make algorithms more efficient and even solve mathematical problems that are unsolvable otherwise. According to some mathematicians, they may in fact be considered the bedrock of a "fundamental source of mathematical truth independent of, but supplementary to, the standard axiomatic structure of mathematics."

It is important, however, to understand the breadth as well as the limitations of using random numbers when it comes to solving mathematical

DOI: 10.1201/9781003641520-20

and computational problems. As elements of algorithm design and for mathematics, random numbers can play three related but distinct roles:

- Speed up computations to provide the most practical solution as well as an unquestionably accurate one.

- Speed up computations, albeit at the cost of introducing a quantifiable probability of doubt attached to that solution. The error probability can, however, be reduced as much as desired.

- Solve problems that we are currently unable to solve using any other method. In these cases, we simply don't know if we will ever have a better option than to utilize random numbers.

We will now examine several examples to appreciate each of these roles, and to recognize random numbers as more than just a tool for simulation, gaming, and polling, but also for solving complex problems.

With the insertion of randomness, a deterministic algorithm is converted into a randomized algorithm. Randomized algorithms are applied to problems in number theory, algebra, pattern matching, sorting, searching, computational geometry, graph theory, data structure maintenance, distributed computing.

To begin with, let's look at an example where random numbers allow us to employ a solution that is both the most practical and unquestionably accurate: the well-known task of sorting a list of numbers (or alphanumeric entries) in ascending order. Computationally, the easiest to understand sorting algorithm is the so-called Bubblesort. A pointer advances through the list and whenever the indicated number is smaller than the preceding one, the two swap places. As soon as the pointer reaches the end of the list, the process starts over again from the top. This is repeated until an entire run-through takes place without any more swaps; at that point the list is sorted with the smallest on top, the largest at the bottom. The name Bubblesort derives from the fact that in the course of successive runs, the small numbers eventually rise to the top, like bubbles in a glass of water.

Unfortunately, Bubblesort is very inefficient. A list that contains n entries may require n swaps at the first run-through, and there could be as many as n runs. Hence, the algorithm may take of the order of n^2 operations.[1] With

---

[1] 'Of the order of n^2' is indicated by mathematicians as $O(n^2)$. In the worst case, if the initial list were totally ordered but in the wrong way, from largest to smallest, there would be $(n-1) + (n-2) + (n-3) + ... + 1$ swaps. This adds to $(n-1) (n-2)/2$ which is $O(n^2)$.

FIGURE 15.1 Randomness, like a *deus ex machina*, can rescue deterministic algorithms by unjamming their worst-case inefficiencies. © ChatGPT.

1,000 entries, there could be 1 million operations. When the length of the list doubles, becoming $2n$, the swaps could quadruple: $(2n)^2 = 4n^2$.

An improvement over the Bubblesort was sorely needed and it came from an unlikely corner. In 1959, the British student Tony Hoare, a scholar of classics and philosophy, was working at Moscow State University on machine translation. In the course of his work, he had to sort the words of Russian sentences and then look them up in a Russian–English dictionary that was stored on magnetic tape. To facilitate the task of putting the words in alphabetic order, he developed a code for a sorting algorithm. When it became known that it was quicker than the Bubblesort, it acquired the name Quicksort. (Hoare later left classics to become professor of computer science at Oxford and principal researcher at Microsoft.)

In Quicksort, one entry, called the pivot, is chosen and compared one by one with the other entries on the list. If an entry to the left of the pivot is greater than the pivot, it is moved to the pivot's right. If an entry on the right is smaller than the pivot, it is moved to its left. All other entries remain in place. After the first run, the pivot is in its correct position, with one sublist containing all the smaller numbers on the left, and another sublist containing all the larger numbers on its right. Then the same procedure is performed on the two sublists: pivots are chosen and the entries are moved to the right, or to the left, or remain in place. Now the pivots of the two sublists are also in their correct positions on the list. And the procedure is repeated with the four sub-sublists, and then with the eight sub-sub-sublists, ... until all entries are in their correct positions. Quicksort belongs to the class of *Divide and Conquer* algorithms because it divides the list into successively smaller lists which are then conquered, that is, ordered in turn.

How quick is Quicksort? Let us assume that on average the chosen pivot is approximately in the middle of the list or of the sublist or of the sub-sublist...; then the two 'offspring' contain more or less half of the 'parent' list. On each run all n entries are compared with the respective pivots to see if they need to be moved. That is an $O(n)$ operation. And at the end of each run, at least one entry has found its correct place in the list. How many runs are required?

Well, one keeps splitting the list roughly in half until each piece is in its correct position. The number of splits needed to do this is about the same as the number of times one can halve the list before getting to single items. If we call the number of splits r, then after r splits, we have:

$$\frac{n}{2^r} = 1 \quad \text{or} \quad 2^r = n$$

Taking the logarithm (base 2) of both sides, we have,

$$r = \log_2(n)$$

Hence, the number of splits (or runs) is proportional to the logarithm of the list size.

So, there are $\log_2 n$ runs, and at each run n comparisons are performed, hence Quicksort's average runtime is of the order of $n \cdot \log_2(n)$. This is much faster than $O(n^2)$ that the Bubblesort required. Since $\log_2(1{,}000) = 9.965...$, Quicksort would require the order of $1{,}000 \log_2(1{,}000) \sim 10{,}000$ operations to put a list of 1,000 entries into order. A list of 2,000 entries would require about 2,000 $\log_2(2,000) \sim 22,000$ operations. Bubblesort, as you recall, would have required n^2, that is, millions of operations.

What does this have to do with random numbers? Well, recall that I wrote above "the pivot, is chosen…" But the algorithm cannot just 'choose' on its own; it must be told what to choose. And this is where random numbers come in. A random number between 0 and $n - 1$ is generated and indicates the first pivot; likewise, random numbers indicate the pivots in the sublists. How these random numbers are generated will be the subject matter of Part V of the book. For the time being, the important thing to note is that the use of random numbers improves the average time complexity of the sorting procedure from $O(n^2)$ with Bubblesort, to $O(n\log_2 n)$ with Quicksort.

❧

The Quicksort algorithm made use of random numbers to order a list perfectly in a much shorter running time. Now, we give an example of an algorithm that also uses random numbers to speed up computational time, but the result is not perfect: there is a minute chance of the result being wrong.

Prime numbers, that is, the integers that can only be divided by themselves and by one, are considered the basic building blocks of higher arithmetic, a.k.a. number theory. One problem of number theory is verifying whether a particular number is prime, its primality. Carl Friedrich Gauss, the nineteenth-century 'prince of mathematics,' considered the study of higher arithmetic, and thus prime numbers, as fundamental to all of mathematics. He expressed his belief succinctly in his famous dictum "mathematics is the queen of the sciences – and number theory is the queen of mathematics."

But theory is one thing, usefulness another. In the early twentieth century, number theorists believed that their field of study, while important for mathematics itself, had absolutely no worthwhile applications. This would change, first during World War II when the Germans used number theory to encrypt messages with the Enigma machine, and the Allies, particularly the British at Bletchley Park, used number theory to decrypt these messages, and again half a century later with the advent of the internet and online trading. To allow electronic transactions to be conducted securely and email messages sent privately, ever more advanced encryption methods had to be developed. It quickly turned out that the answer could be found using number theory, vastly improving its estimation in the eyes of even the most pragmatic.

The secret to the use of prime numbers in encryption is that while it is easy to multiply prime numbers, it is much more difficult to factorize a composite number. Take the two prime numbers 1,697 and 5,147. It is easy to multiply them to get the composite number 8,734,459. On the other hand, had you been presented with the number 8,734,459, you would have been hard pressed to find the two prime factors. That's the secret of encryption…with some bells and whistles added on.

Of course, the larger the prime numbers that are used, the more difficult it is to factorize their product and the better is the encryption. And when I say *large* prime numbers, I mean *large, large*; prime numbers 500 digits or more in length are nothing out of the ordinary. So, how does one verify whether an integer is prime or composite?

For small numbers, the verification method is straightforward. Simply divide the number by successive primes up to $\sqrt{n}$ to see if you hit on a factor that does not leave a remainder. Why $\sqrt{n}$? Because for any factor larger than that there would exist the other factor that is smaller and which would have already been found. So, the runtime of the algorithm will be of the order $O(\sqrt{n})$. For reasonably small numbers, like 8,734,463, the two factors 1,847 and 4,729 will be found very quickly. (Re-running the algorithm with these two factors produces no divisors, hence they are prime.)

How about a number like 7,248,103,372,958,048,761? The algorithm must try out all numbers up to the square root of that integer, that is, up to about 2.7 billion. My laptop was able to factor it in a bit over eight seconds.[2] But then again, this quintillion-integer has only a measly 19 digits. What about searching for factors of integers that are several hundreds of decimal digits long? In early 2020, an international team of computer scientists reported that they managed to factor a 250-digit number in a few months… with the help of a network of tens of thousands of machines worldwide.[3] And even 250 digits are peanuts in cryptography.

So, for all practical purposes, factoring large integers is infeasible. Luckily, for purposes of encryption it suffices to know that an integer is prime; one does not need to find the actual factors. (The attackers are the ones who would need to find the factors… but they are not our concern.) And there is another easement: one may content oneself with the

---

[2] In fact, the factors are 3,203,475,413 and 2,262,574,997.
[3] The computations took the equivalent of 2,700 years on a single central processing unit (CPU).

knowledge that an integer is most probably prime; absolute certainty is not required. So, the task is to develop a primality test, that is, an efficient algorithm that confirms with a high probability that a specific integer is prime. Fermat's little theorem – the name given to distinguish it from the much more famous last theorem – was a first step.

In the following two paragraphs, I've provided a brief introduction to the mechanics of this theorem and its weaknesses, but don't fear if its inner workings feel oblique and you may skip the technical details. The critical takeaway for our purposes is that the theorem can suggest that a given number is prime, albeit only to within a probability; unfortunately, there is no method to quantify that probability.

The French mathematician Pierre de Fermat (1607–1665) provided a test, albeit not for the primality of numbers, but for their compositeness.[4] His little theorem proved that if a number P is prime, then, for any randomly chosen number A that is smaller than P, and that has no common divisors with P, A^{P-1} is equal to *one* plus a multiple of P, or $1 + kP$. In modular arithmetic, this can, of course, be written as,

$$A^{P-1} = 1 \bmod (P)$$

This requires context. If A^{P-1} is unequal to $1 + kP$, then, by Fermat's little theorem, P is composite for sure. On the other hand, if it is equal to $1 + kP$, it may be prime but need not be. In this case, P is called pseudo-prime. To gain additional confidence in whether P is truly prime or pseudo-prime, one runs the test with several more randomly chosen A's. If they all result in $1 + kP$, we can be more confident of the number's primality...but still cannot be certain because there are the Carmichael numbers which, though very rare but nevertheless infinite in number, are composite numbers that fool the little theorem with any choice of A.

So, primality tests allow us to distinguish composite numbers from numbers which are probably prime. How probable? Well, the error is measured by the probability that a composite number is erroneously declared prime. Regrettably, the little theorem does not provide an estimate for that

---

[4]   For a proof of Fermat's little theorem, see any textbook on number theory.

probability; we need another method, which we now have in what is known as the Miller–Rabin test.

In 1976, the American computer scientist Gary L. Miller found a primality test that would identify prime numbers unequivocally. But there was a snag: Miller's proof depended on the correctness of the notorious Riemann hypothesis, the most famous unproven mathematical hypothesis of all times. Hence Miller's test cannot be considered conclusive. Five years later, in 1980, the Israeli computer scientist Michael Rabin developed a variant of Miller's test which did not rely on Riemann hypothesis. But now there was another snag: Rabin's test was equivocal; it only identified prime numbers with a certain probability. However, in contrast to Fermat's test, the probability could be quantified.

The Miller–Rabin test first reduces the number P, that is suspected of being prime, by dividing $P - 1$ by 2 as often as possible, let's say s times. Hence, $P - 1 = 2^s \times U$, where U is odd. The test relies, as does the Fermat test, on a number A, randomly chosen between 0 and P. If P is, in fact, prime, one of the two following conditions must be fulfilled: either $A^U = 1 \bmod P$ or $A^{2^r \times U} = -1 \bmod P$ for some r, $0 \leq r < s$.

Unfortunately, there are numbers A that fulfill one of the requirements even though P is composite. Rabin was able to show, however, that this can occur at most for one quarter of the numbers A. So, the probability that the test erroneously identifies a number as prime, even though it is composite, does not exceed ¼. And this is the crucial point: by performing the test with several randomly chosen A, the probability of an erroneous result can be reduced each time by three-fourths. After running the test with, say, 10 random As, the probability of erroneously ascertaining a composite number as prime is only about 0.0001%, with 20 As it is reduced to 0.0000000001%.[5]

Another use of random numbers in algorithms is polynomial identity testing. Again, their introduction into the algorithm allows us to obtain a practical solution, though it nevertheless comes with a miniscule, but non-zero, probability of error.

---

[5] In 2002, M. Agrawal, N. Kayal, and N. Saxena published a deterministic algorithm, that is, one that does not require random numbers to ascertain primality probabilistically. However, it is orders of magnitude slower than the Miller–Rabin test (see Chapter 19).

The question is whether two given polynomials are identical, simply expressed in different ways, or whether they are indeed different. We can easily verify that the two-degree polynomial with two variables $(a + b)(a - b)$ is identical to $a^2 - b^2$ by expanding all products and checking whether the monomials are identical: $(a + b)(a - b) = a^2 - ab + ab - b^2 = a^2 - b^2$.

But how about $ab^2 + c^3(a^2 - b)(b^3 - c^2)$ and $a(b^2(1 + abc^3) - ac^5) - bc^3(b^3 - c^2)$? Are they identical or different? The more variables the polynomials contain, and the higher their degrees, the more difficult the problem becomes...until it is quite intractable. In fact, as a polynomial's degree, d, and number of variables, v, grows, the number of monomials increases as $(v + d)!/d!v!$, which becomes exponentially large.[6] A polynomial of degree 10, with 20 variables, would expand into about 30 million monomials.

A simple test for polynomial identity would simply insert numerical values into the polynomials, say $a = 2$, and $b = 3$, to verify that $(a + b)$ $(a - b)$ gives the same result as $a^2 - b^2$, namely, in both cases, -5. But this is not good enough. With these numerical values, another polynomial, $-a^3 + b$ for example, also results in -5, though it is not at all identical to $(a + b)$ $(a - b)$. In general, to verify whether two polynomials, $P(x)$ and $Q(x)$, are identical, we form the new polynomial $P(x) - Q(x)$ and check whether the answer is zero. To simplify the following exposition, we restrict ourselves to polynomials in one variable. Take $P(x) = (x + 1)(x - 2)(x + 3)$ and $Q(x) = x^3 + 6$. Is $P(x) - Q(x)$ equal to zero? (Spoiler alert: it is not!)

We randomly choose a numerical value for x, say 4, and insert it into the two polynomials. In both cases we get the result 70, which means that $P(4) - Q(4) = 0$. Does that mean that the two polynomials are identical? Well, a value that makes a polynomial become zero is called a root; and it so happens that the number 4, supposedly chosen at random, is a root of $P(x) - Q(x)$. The conclusion is that accidentally hitting on a root when choosing a random number for x may lead us to assume, erroneously, that the two polynomials are identical, even though all that occurred was that the two polynomials just happened to be equal at that particular value of x.

Taking a cue from the Miller–Rabin algorithm for prime numbers, we could simply re-test the two polynomials with several values of x. If just one of them gives different results, we have conclusive evidence that the polynomials differ. Let us take $x = 3$. For $P(3)$ we get 24, for $Q(3)$ we get 33. Hence $P(x)$ is not identical to $Q(x)$.

---

[6] $\binom{v + d}{d} = (v + 1)(v + 2)... (v + d)/(1 \times 2 \times 3 \times ... \times d)$.

The converse is not true, however: even if one or more xs pass the tests, we do not have conclusive evidence that the polynomials are identical. For all we know, all the xs could be roots of $P(x) - Q(x)$.

What is required is a bound on the probability of error. What is the probability that the two polynomials differ if we choose a value *at random*, insert it into $P(x) - Q(x)$, and get zero?

The answer to this question was given in the late 1970s independently by NYU's Jack Schwartz, Richard Zippel, then at MIT, and Richard DeMillio and Richard Lipton of the Georgia Institute of Technology. They found that choosing a random number that, just by chance, produces the same result for both polynomials, that is, that it is a root of $P(x) - Q(x)$, is very rare. The fundamental theorem of algebra, known since the early seventeenth century and proved in the eighteenth, states that a polynomial of degree D has at most D real roots. Hence, there can exist at most D values of x which would make $P(x)$ equal to $Q(x)$.

The four computer scientists proved that if x is randomly chosen from a pool of numbers that contains S elements, the probability of hitting on a root is at most D/S. Hence, the probability of choosing a root of $P(x) - Q(x)$ from among the infinitely many real numbers is close to zero. Whenever the algorithm is re-run with another randomly chosen x, the probability is reduced further.

The second role laid out at the beginning of this chapter is pervasive and important enough that it is worth exploring again. Randomized algorithms come to the rescue, providing a speedy if not quite absolute solution, in the case of the multiplication of mathematical objects called square matrices. Square matrices are arrays of numbers arranged in n rows and n columns. There are also rectangular matrices which have m columns and n rows, or the so-called vectors which have just one column and n rows (or n columns and one row). The multiplication of two $n \times n$ matrices A and B – devised in 1812 by the French mathematician Jacques Marie Philippe Binet – is an involved procedure that produces another $n \times n$ matrix C. It consists of entering into the ith row and jth column of matrix C the sum of the products from cells in the ith row of A and the jth column of B ($c_{ij} = a_{i1}b_{1j}$ $+ a_{i2}b_{2j} + a_{i3}b_{3j} +...$). Hence, the multiplication of an $n \times n$ matrix with another $n \times n$ matrix produces another $n \times n$ matrix; the multiplication of an $n \times n$ matrix with a vector produces a vector.

A bit confusing? Yes, and that is why it is quite difficult to verify whether a matrix C is, in fact, the product of A and B. Even computer algorithms can take very long times when n is large: the computation for each of C's cells involves n multiplications, and since the matrix contains n^2 cells, the total number of steps that the algorithm performs is of the order of n^3.

I will now describe a faster algorithm that helps with the verification. Readers less interested in mathematical technicalities, may skip them. The important thing to note is that the sped-up algorithm allows verification of whether matrix C is, in fact, the product of A and B, albeit with a probability of error that can, however, be made arbitrarily small.

&

In 1977, the Latvian computer scientist Rūsiņš Mārtiņš Freivald came up with an ingenious trick. To understand it, note that the multiplication of an $n \times n$ matrix, say C, with a vector, say r (which is, in effect, an $n \times 1$ matrix), produces another vector, $C \times r$. The computation of each cell of $C \times r$ requires n multiplications, and since the vector has n cells, multiplying a matrix with a vector requires only n^2 steps. So, if $A \times B$ is equal to C, then $A \times B \times r$ must be equal to $C \times r$.

What Freivald suggested was, first, to create a vector, r, of n randomly generated numbers and then multiply matrix B with vector r. (For simplicity, these numbers are usually just random bits, i.e., zeros or ones.) This multiplication produces a vector, hence only requires n^2 multiplications. Next, the multiplication of matrix A with the vector $B \times r$ and the multiplication of matrix C with the vector r – which produce vectors – also require only n^2 steps. So, instead of multiplying two matrices with each other, Freivald's procedure requires the multiplication of three matrices with a vector; the total number of steps that the algorithm performs is of the order of n^2. Hence, the computational complexity has been reduced from $O(n^3)$ to $O(n^2)$.[7]

The next step requires verification whether $A \times B \times r = C \times r$. But once again, the speed-up comes at a price. If it turns out that $A \times B \times r$ is not equal to $C \times r$, then the algorithm has provided conclusive proof that C is not the product of A and B. So far, so good. But the converse is not true: even if $A \times B \times r$ turns out to be equal to $C \times r$, the algorithm has *not* provided conclusive proof that C is

---

[7] The number of steps in the verification grows only linearly with n, that is, is of the order $O(n)$, and is therefore ignored here.

the product of A and B. For it could be that $C \times r = A \times B \times r$ by coincidence. How so?

Let's denote $A \times B - C$ by D. If $A \times B$ is equal to C, then all entries of D must be zero. If just one entry is non-zero, $A \times B$ is not equal to C. Here we can bring back the vector r to decrease the number of steps needed for this computation. The question becomes whether all entries of $D \times r$ can be zero by coincidence, even though not all entries of D are zero? We answer the question, not to give a proof but by an example.

Let's assume that all entries in D are zero, except for the entry in row p column q. By definition, the entry in the qth cell of vector $D \times r$ is $d_{p1}r_1 + d_{p2}r_2 + \ldots + d_{pq}r_q + \ldots + d_{pn}r_n$ which corresponds to $0r_1 + 0r_2 + \ldots + d_{pq}r_q + \ldots + 0r_n$. Now, if the random bit r_q happens to be zero, then all of the entries in the vector $D \times r$ are zero, including the one in qth cell…even though D is unequal to zero. The algorithm gave an erroneous result!

Fortunately, we can again quantify the probability of the error. Since the values of r_i are randomly generated bits, the probability of picking a zero for r_q is ½. And now it's *déjà vu* all over again: by running the algorithm with different randomly chosen 0/1-values for the vector components r_i, the error probability is reduced at each run, that is, it is reduced to $(1/2)^k$ after k runs. The runtime is $O(kn^2)$ which, for large n, is much less than $O(n^3)$.

The search for the so-called *minimum cut* in a network is an example of how an algorithm uses random numbers to solve a problem that can currently not be solved in any other way, except by brute force. To illustrate, let us consider a network of streets in a city.[8] To contain riots that have broken out all over, the sheriff wants to block streets, such that the city is separated into two disconnected parts. Since manpower is limited, the number of roadblocks must be minimized. Which streets should be blocked?

This question was the subject of a 1995 PhD thesis at Stanford University by David Karger, later professor of computer science at MIT. His idea was to choose an edge (that's the word we'll use instead of street) and contract it, such that the two nodes (that's the word we'll use for the intersections at

---

[8] The algorithm is also applicable for networks like Facebook, the internet, railway lines, organized crime, terror networks, telephone networks, the transmission of diseases, dissemination of information, etc.

each end of a street) are concatenated, that is, fused together. As a conse-quence, all edges that connect these two concatenated nodes via other nodes also disappear. Then randomly choose another edge and fuse the nodes. Repeat this task over and over again, until only two nodes are left. These two nodes, each of them being a concatenation of many nodes, represent the two parts of the network. They are connected by the last remaining edges. Once these edges are cut, the two parts of the network are separated.

It is suspected with high probability that these last remaining edges represent the minimum cut, that is, they are the smallest number of roads that the sheriff needs to block. Why is that? Well, there are many edges in a network, but only a few that represent a minimum cut. Hence, when edges that are to be contracted are chosen at random, they are most likely not from among the ones that potentially belong to a minimum cut.[9]

However, a first analysis of the probabilities does not look good. True, when randomly picking the first edge – that is, concatenating the first two nodes – the probability that it does not belong to the minimum cut (which is what we are hoping for) is $^{n-2}/_n$, that is, for a network with 10 nodes it is 80%. But the more nodes are concatenated, the larger the probability of catching an edge that actually does lie in the minimum cut (which is what we want to avoid). The probability of getting down to just two nodes, without picking any of the minimum-cut edges, becomes $^2/_{n(n-1)}$. For $n = 10$ this yields a humbling 2.222…%. In other words, there's a 97.888…% chance of randomly picking at least one edge that should not have been picked.

By now we know what to do in order to get better probabilities: run the loop again…and again…and again…and then chose the best cut. The probability of not finding the minimum cut after running the loop T times is $(1- ^2/_{n(n-1)})^T$, which, for a network with 100 nodes, and 20,000 runs is only 1.758…%. In other words, the minimum cut is found with a probability of 98.241…%.

Note that the Karger algorithm requires vast amounts of random numbers. To concatenate 100 nodes to just 2, requires 98 random numbers, and running the loop 20,000 times means that altogether nearly 2 million random numbers are needed.

---

9   The *MinCut* problem can be solved by brute force by iterating over all pairs of vertices; hence the algorithm will run in $O(n^2)$. The Stoer–Wagner algorithm produces the true minimum cut and runs in polynomial time.

The examples in this chapter explore three roles that random numbers can play in solving problems of computation and mathematics. Though some such problems can be solved, at least theoretically, with brute force (by testing all of the myriad possibilities), or with cumbersome algorithms (e.g., the Bubblesort), making use of random numbers allows algorithms to become much more efficient and faster. Some randomized algorithms (like Quicksort) produce the desired solution. Others (primality testing, matrix multiplication) produce acceptable solutions, albeit tainted by a minute probability of being incorrect. Still others (e.g., the minimum cut of a network) can currently only be solved – also with a probability of being wrong – with random numbers.

In this chapter, we showed how problems can be solved by incorporating random numbers into algorithms. But the very fact that we can carry a difficult problem from the traditional, deterministic world into one built on random numbers begs a question: might we, in some cases, be able to carry that problem back in a new and improved form, combining the best of both worlds?

In the next chapter, we make a U-turn and try to pull random numbers out again, without sacrificing the improvements that randomness provided. Derandomized algorithms are algorithms that originally relied on random choices to achieve their results but have been modified to eliminate or reduce the use of randomness, while still maintaining their efficiency and correctness. Hence, derandomization techniques transform probabilistic (randomized) algorithms into deterministic ones which produce the same outcome, at roughly the same efficiency, without the reliance on random inputs.[10]

---

[10] However, it is not quite true that randomization (or brute force) are the only options: efficient methods now exist (de-randomized algorithms, see next chapter), running in near-linear time, although they are less intuitive than Karger's.

CHAPTER 16

Making a U-Turn

Derandomization

As we saw in the previous chapter, random numbers can provide extraordinary shortcuts to seemingly intractable computational problems. By allowing a minute probability of error, randomized algorithms – that is algorithms that rely on random numbers or random bits – take tasks that would keep a deterministic algorithm busy for many years and speed them up to a manageable timeframe. Thus, by inserting random numbers into the process, tasks that, for all practical purposes, would be unsolvable if they were purely deterministic, or computational problems that would be wholly intractable, can be solved efficiently.

This begs a question: if the introduction of random numbers can enhance an algorithm's performance, are there methods to remove them without reverting the algorithm to its original, unmanageable timeframe, and without introducing a margin of error?

Then answer is a qualified yes…

So, let us now make a veritable U-turn: after having introduced randomness into algorithms in the previous chapter, we now discuss how randomness can be removed from them. Sometimes it takes only the addition of a few routines to remove it.

DOI: 10.1201/9781003641520-21

Let us remind ourselves why randomness was introduced in the first place. It may be that, historically, no deterministic algorithms were known that were able to solve certain problems. Or the deterministic algorithms would require impossibly long run times. Randomization may have been the only known way to solve certain problems.

Now, one may legitimately ask why get rid of randomness if the results of randomized algorithms are acceptable. There are several reasons. For one, random numbers are regarded a scarce resource, just like computing time and memory space. Or one may distrust the random number generator. (We will discuss reasons for the distrust in Chapter 18.). At other times, one may want to re-run an algorithm with the same input – to debug the code, for example – but with constantly changing random numbers, outputs are non-reproducible. (That particular problem can actually be avoided by re-running the algorithm with the same sequence of random numbers.) Rarely, even astronomically small probabilities of failure are unacceptable. Occasionally, new insights may be gained. Finally, there may be psychological reasons: a common dislike of uncertainty. In general, if at all possible, it is better to avoid randomness.

So, why would one not head straight for a deterministic algorithm instead of derandomizing a randomized algorithm? Well, developing randomized algorithms for a particular problem provides programmers with knowledge, experience, skills, and tricks. To devise a deterministic version that solves that problem, or speeds up runtime, programmers can make use of the expertise that they gained by developing and working with the randomized predecessor.

Truth be told, even randomized algorithms are deterministic, in principle. As pointed out before, only physical phenomena – like coin throws, atmospheric noise, quantum phenomena, … – can produce truly random numbers. Computers are deterministic, the best that algorithms can do, as we shall see in Part V of this book, is to generate *pseudo-random* numbers. The sole truly random item in a pseudo-random number sequence is the initial number, the so-called seed. Once the seed has been chosen, all other pseudo-random numbers are predetermined

❧

To illustrate derandomization, let us have another look at the *Quicksort* algorithm that was discussed previously. A significant improvement over *Bubblesort*, this *divide and conquer* algorithm sorts a list of alphanumeric items by choosing a pivot element and arranging the smaller entries to the pivot's left, the larger entries to its right, and then repeating the process for the sublists on the left and on the right, until the entire list is ordered.

The crucial point here is how the pivot element is chosen. A naïve version of *Quicksort* begins the process by choosing either the leftmost or the rightmost element of the list (and later of the sublists) as pivots. This is a deterministic algorithm which has the unfortunate disadvantage, however, that it is very inefficient on lists that are already completely or partially ordered. In such cases, the algorithm may take up to n^2 steps to put n elements in order. A more efficient choice of pivot elements is to pick them at random which, as we discussed, reduces the average runtime to $n \cdot \log_2(n)$.

Now to the U-turn, if one wants to get rid of randomness, for whatever reason, the choice of pivots can be made deterministic again, though in a different way than in the naïve approach. Recall that *Quicksort* performs worst if the pivot is chosen at the beginning or at the end of a list that is already ordered or partially ordered. The best performance, on the other hand, occurs when the chosen pivots divide the lists into approximately equal sublists. Hence, the best method would be to take the lists' or the sublists' median each time and arrange all entries to its left and to its right.

To find the median is computationally intensive, however.[1] Hence, one makes do by picking a small sample and taking *its* median. A common strategy is to inspect elements at the list's beginning, middle, and end, and to choose the median of the three. Though this does not guarantee anything, it is very likely to give good performance. Best of all, there is no longer any randomness in the algorithm; it has been derandomized!

For a more sophisticated example, let us recall the sheriff who had to separate his city into two disconnected parts, *A* and *B*, by blocking streets. To save on workforce, he strove to block the fewest roads. This problem is called *MinCut*, and there are algorithms that solve it within reasonable time: as the crossroads increase, runtimes increase no more than by a polynomial in the number of crossroads.[2]

Now let us discuss a related but different problem. The police union wants to increase jobs and therefore the sheriff strives to maximize the streets that must be blocked. This is called the *MaxCut* problem and it is

---

[1]   It would itself require sorting the list or employing another algorithm which runs in linear time.
[2]   In Chapter 15, we also mentioned the randomized Karger algorithm which is even more efficient.

quite a different animal. In contrast to the *MinCut* problem, it is not known whether the *MaxCut* problem can be solved in polynomial time.[3]

One obvious, but impossible, way to find *MaxCut* would be to run the algorithm…and to re-run it…and to re-run it…for all possible combinations of assignments to groups *A* and *B*. Such an enumeration entails 2^N runs. The advantage would be that since all possible assignments of the nodes to groups *A* or *B* would be included, we could be certain to hit upon at least one that gives us the true *MaxCut*. The disadvantage, say, the absolute obstacle to such an enumeration is, of course, that the algorithm would have to be run 2^N times. Even for a moderately large network with, say, 40 nodes, this would mean over a trillion runs.

So, since the true *MaxCut* of a large network can, for all practical purposes, not be found, one usually makes do with algorithms that provide 'good enough' cuts, that are guaranteed to block at least half of all streets.[4] We first describe a randomized algorithm, and then two derandomized algorithms which do just that. In keeping with the language of graph theory, to which the *MaxCut* problem belongs, we shall again denote the street segments as 'edges' and the points where edges cross each other as 'nodes' (or as vertices' in graph theory).

First, let us review the randomized algorithm. Nodes are numbered from 1 to *n* and we proceed down the list to assign the nodes to groups *A* and *B*. How do we do it? By flipping coins, or rather by producing pseudo-random bits. If the bit is zero, the node is assigned to *A*, if it is one it is assigned to *B*.

Consider an edge together with the two nodes at each end, and the random bits that determine the nodes' fate. If the bits are equal, the nodes lie in the same group and the edge is not cut. If the bits are different (one bit is zero, the other is one) the nodes lie in different groups and the edge is cut. The probability of the two independent random bits being different is ½; hence one half of the edges are expected to lie on the cut.

With *N* nodes in the network, the randomized algorithm requires one random bit for each node, that is, a series of *N* zeros and ones. Of the 2^N possible series of bits, the algorithm chooses just one at random.

And now on to the derandomized algorithms, we will describe two examples.

---

[3] In fact, the *MaxCut* problem belongs to the class of *NP hard* problem which – as the name says – are hard. See Chapter 19 for more on *NP* problems.

[4] The algorithm is usually described as cutting at least half as many roads as the best (but unknown) *MaxCut*; hence it is an 'approximation' algorithm.

The first deterministic algorithm belongs to the class of so-called greedy algorithms. Nodes are again numbered from 1 to n and we start by assigning node 1 to group A, and node 2 to group B. Then, proceeding down the list, each node is assigned either to A or to B. This time the assignment is not random, however, but according to which assignment produces more crossings. If, say, nine edges lead towards the node under consideration, five originating in group A, two in group B, and two are not yet defined, then the algorithm assigns this node to B, so that five edges will be added to the cut. This means that in the end at least half of all edges will be on the cut between A and B. (The algorithm is called greedy because it always looks for what is the best option at that very moment.)

The result may not be anywhere near the *MaxCut*; since greedy algorithms are short-sighted, they do not consider whether it would have been advantageous for the overall picture to assign a node to the other group. But at least we obtained a 'good enough' cut that comprises at a minimum half the edges.

Recall that a full enumeration would have required running the algorithm for all possible combinations of assignments to groups A and B, while the randomized algorithm described earlier, chose just one of the 2^N series at random.

Another method of derandomizing the algorithm is to do something in between: enumerate more than just one series of random numbers but only a tiny slice of the 2^N series that the full enumeration would have required. The algorithm will most probably not stumble upon the true *MaxCut*; but as I will show, even such a partial enumeration will find a cut that contains at least half the edges.

So, how can one reduce, say, a trillion sequences of 40 random bits to half a dozen? Recall that just a few paragraphs ago, we specified that the probability of the two independent random bits being equal or unequal is ½. Hence, the bits that determine whether the two ends of an edge come to lie in the same or in different groups need to be independent only *of each other*, that is, they must be pairwise independent; they do not need to be globally independent of the other $N - 2$ bits that determine the remaining nodes' group affiliation. If we can generate pairwise independent random variables, using less randomness than it takes to generate completely independent random variables, then the derandomized version of the randomized algorithm described before would be more efficient. We would only enumerate (run the algorithm) on a subset of the possible series.

Let us describe how to generate pairwise independent triples of random bits. First, we create sets of three random bits. They could be any of the following: (0,0,0), (0,0,1), (0,1,0), (0,1,1), (1,0,0), (1,0,1), (1,1,0), (1,1,0). The random bits in these eight sets are completely independent since they cover all possible sets of three bits.

On the other hand, let us now create sets of three bits, only two of which are random, while the third is determined by the first two in the following manner: if the two first bits are identical, the third will be zero; if they are different, it will be one. There are only four possible sets: (0,0,0), (0,1,1), (1,0,1), (1,1,0). Though the three bits are not independent – the third is determined by the first two – we know that the first and second bits are pairwise independent, and so are the first and third, and the second and third. Thus, by using only *two* independent random bits we were able to generate *three* pairwise independent random bits. As a consequence, we would need to run the algorithm only four times instead of eight times. The question is by how much one can, in general, reduce the number of totally independent random bits if only pairwise independence is required.

The good news is that there is a way to generate N pairwise independent bits from only $\log(N+1)$ truly random bits. And if the algorithm requires only logarithmically many random bits, then one can make it deterministic by exhaustive search. For a network with 40 nodes, for example, we would require only 6 truly random bits, and instead of 2^{40} runs that a total enumeration would require, the partial enumeration would require only 2^{6}.[5]

One way of generating many pairwise independent bits from just a few truly random bits is with the bitwise XOR operator ('exclusive or' denoted by $\oplus$). The variable $c = a \oplus b$ is defined, as earlier: if a and b are identical, c will be zero; if they are different, c will be one. Indeed, that is what we did when we reduced three globally independent bits to two pairwise independent bits and reduced eight sets of bits to just four.

To create a variable that depends on more than two random bits but is pairwise independent of all others, the XOR operation is performed sequentially: $d = (c \oplus (a \oplus b))$. Effectively, the XOR operation returns the parity of the independent variables, that is, if the number of ones in a, b, and c is odd, it is one; otherwise, it is zero (010 gives one, 011 gives zero).

In order to create $2^r - 1$ series of random bits that are pairwise independent of each other, XOR operations are performed with every

---

[5] Actually, 6 truly random bits would allow the generation of pairwise independent bits for a network with 63 nodes.

combination of r truly random bits. If, for example, we have generated four series of truly independent random bits, r_1, r_2, r_3, r_4, we can create 11 additional series of bits, to give a total of $2^4 - 1$ pairwise independent series of random bits: we create

$$r_5 = r_1 \oplus r_2, \quad r_6 = r_1 \oplus r_3, \quad r_7 = r_1 \oplus r_4, \quad r_8 = r_2 \oplus r_3,$$
$$r_9 = r_2 \oplus r_4, \quad r_{10} = r_3 \oplus r_4, \quad r_{11} = r_1 \oplus r_2 \oplus r_3, \quad r_{12} = r_1 \oplus r_2 \oplus r_4,$$
$$r_{13} = r_1 \oplus r_3 \oplus r_4, \quad r_{14} = r_2 \oplus r_3 \oplus r_4, \quad r_{15} = r_1 \oplus r_2 \oplus r_3 \oplus r_4.$$

Though only the first four series, r_1, r_2, r_3, r_4, are globally independently random, all 15 are pairwise independently random. Thus, we managed to reduce the required number of random bits from 15 to 4, and the number of runs from 2^{15} (=32,768) to 2^4 (=16). All we have to do is run the algorithm with the 16 possible allocations of zeros and ones that the 4 independent bits r_1, r_2, r_3, r_4 can assume.[6] As before, the algorithm is guaranteed to give cuts that contain at least half the edges.

Once the algorithm is run on the reduced series of bits, it has been derandomized; we have effectively removed all randomness and the algorithm is now deterministic. The number of runs required to derandomize the algorithm for networks with N nodes is reduced from 2^N to $2^{\log(N+1)}$. To compute the runtime, the number of runs must be multiplied by the time it takes to run the core routine that counts the edges to A and to B; altogether this corresponds to a complexity of $O(N)$.

All this begs the question: can any randomized algorithms be derandomized? The trivial answer is yes, as was pointed out earlier, since the algorithm could simply be run and re-run with all possible combinations of bits. There would be no randomness left at all. However, if the original randomized algorithm utilizes r random bits, the derandomized algorithm would have to run with each one of the 2^r possible bit strings. Obviously, with large r, this becomes infeasible.

Hence, the related but more relevant question that computer scientists ask themselves is: can any randomized algorithms be derandomized, such that it terminates within reasonable time? By reasonable time is meant a running time that behaves like a polynomial in n, where n stands for the size

---

[6]   Actually only $2^4 - 1$ runs are necessary, since the trivial string of all zeros does not need to be run.

of the problem – for example, the number of digits of the input, the nodes in a graph, the number of cities to be visited by a travelling salesman, etc.

Problems solvable in polynomial time by a deterministic algorithm are denoted by computer scientists as class *P* problems solvable in polynomial time by a randomized algorithm as the class *BPP*. *BPP* stands for *bounded-error probabilistic polynomial time* and refers to the fact that randomized algorithms, though speedy, may provide erroneous answers with a certain probability. As we saw in the previous chapter, this error-probability is bounded, however, and can be made arbitrarily small by running the algorithm repeatedly with different sets of random numbers.

So, we stipulate that the runtime of the core routine increases by no more than a polynomial in the size of the input, that is, it grows as r^d, where r is the input size, and d the degree of the polynomial. After all, this is what is meant by saying that the core algorithm lies in class *P*. In contrast to exponential runtimes, polynomial runtimes are considered manageable with modern computers, since r^d is generally smaller than 2^{r}.[7]

As was pointed out several times before, one can, in principle, derandomize any randomized algorithm simply by running the core routine with all possible combinations of random bits, and enumerating the results. The total running time of such a derandomized algorithm is determined by two factors: the size of the input and r, the number of (pseudo-)random bits that the core algorithm utilizes.

Since the polytime core routine must be run for all combinations of random bits in order to derandomize by enumeration, the number of runs is doubled for each additional bit (once with the bit set to zero, and once with it set to one). Such algorithms, whose runtimes grow like 2^r (exponentially), belong to the class of deterministic algorithms that require exponential time, denoted by *EXP*. Hence, algorithms that derandomize by enumerating all possible combinations of random bits lie in the class *EXP*.

More concisely, the question whether any randomized algorithms can be derandomized boils down to whether every problem from class *BPP* (i.e., every problem that can be solved in polynomial time by a randomized algorithm, albeit with a certain probability of providing an incorrect answer) is also in class *P*. Obviously, class *P* is contained in class *BPP*, since

---

[7]   To wit: say an algorithm requires n^5 steps. When run on a smallish network that contains $n = 40$ nodes, the algorithm performs of the order of 100 million operations. In contrast an algorithm that requires 2^n steps would require of the order of a trillion operations.

we can always randomize algorithms that run in polynomial time. Hence, the real question is: is class *BPP* in class *P*? Can all randomized algorithms be derandomized? The answer to the question – which has become well-known under the catchy motto '*P = BPP*?' – is believed by many computer scientists to be in the affirmative… but so far nobody knows for sure.

Stated as a conjecture, *P=BPP* asserts that one can always get rid of randomness in an efficient manner: if there exists a randomized polytime algorithm for a problem, then there exists an algorithm for the same problem that contains no randomness, and with at most a polynomial blow-up in time. Is the conjecture true? The jury is still out on this one.

∾

If we conjecture that for every *exponential* runtime algorithm there exists a randomized *polytime* counterpart, that is, if *EXP = BPP*, randomness would be a solution even for algorithms that for now have to run in exponential time. If, on top of that, the conjecture *BPP = P* turns out to be correct, that is, if every *randomized* polytime algorithm has a *deterministic* polytime counterpart, randomness would be superfluous. That would be the jackpot: even the algorithms that require exponential time could be run in polynomial time…without randomization.

The computer scientists Russell Impagliazzo and Avi Wigderson hedged at first: "A priori, neither extreme seems likely: there are some problems where randomness seems exponentially helpful, but many hard problems are not susceptible to randomized solutions."

But then they put their money squarely on *BPP = P*. In a seminal paper, they proved that on condition that a suitable pseudo-random number generator exists, *BPP* is indeed equal to *P*. Many experts suspect that eventually someone will be able to prove that such a generator does, in fact, exist – maybe even be able to create one – but so far, none is known.

Postscriptum: There is another aspect to *BPP*. Since randomness can be regarded a scarce resource, just like computer time and memory space, the question is whether randomness can be a substitute for time? If *BPP* lies in *EXP* then one can say that it is possible to get rid of randomness at the price of running the core routine exponentially more times. Vice versa, one can reduce runtime from exponential to polynomial at the price of accepting randomness in the algorithm. If *BPP = P*, then it is a toss-up; algorithms of both classes are polynomial and it depends on which one is more efficient.

∾

Part IV explored the crucial role of random numbers, from their theoretical foundations in pure mathematics to practical applications in cryptography, Monte Carlo simulations, zero-knowledge proofs, and randomized algorithms. We also examined derandomization techniques to reduce reliance on randomness. Moving into Part V, we shift our focus from utilizing random numbers to generating and simulating them. We will explore pseudo-random numbers, the pitfalls of Monte Carlo methods, and the art of extracting randomness from seemingly deterministic sources.

V

**Random Numbers
How Do We Fake Them?**

CHAPTER 17

If You Can't Make It, Fake It

Pseudo-Random Numbers

We've seen how powerful random numbers can be and why they are in such demand. In fact, they are needed so often, for so many processes, and in such enormous quantities that an obvious question arises: why not use computers to produce them?

The answer is that this is impossible. Given the deterministic nature of computing, algorithms can produce only what are known as *pseudo-random* numbers, a concept that we will explore in this chapter. But even pseudo-random numbers are difficult to produce and several software attempts have resulted in notable failures. These challenges underscore the complexity, rarity, and value of high-quality quasi-random numbers.

Indeed, with the advent of electronic computers after World War II, it was not long before these then new-fangled devices were put to use to generate pseudo-random numbers. The earliest endeavor was attempted after the Manhattan Project in Los Alamos had ended, by its towering mathematical intellect, John von Neumann. Von Neumann's achievements are legion, but this time he was off... .by a lot.

He described the algorithm in 1951 in the paper "Various Techniques Used in Connection with Random Digits." Start with some four-digit number, he advised, square it – which usually produces an eight-digit

DOI: 10.1201/9781003641520-23

number – and take the middle four digits as the first random number. Use this number. Then square it, and take the middle four digits to produce the second; and so on, and so on…

And so on, and so on?

Oh no! After only a few iterations, the sequence enters a loop, reproducing the same sequence of numbers over and over again. The problem is that whenever one of the four-digit numbers ends with two zeros, which will occur sooner or later, these zeroes perpetuate themselves. Though the sequence can be made somewhat longer by using, say, ten-digits instead of four, it will eventually enter a loop.

There was another problem, one that will pop up throughout this book: the numbers in the sequence are not independent of each other. Each one is determined by the preceding one, and all of them depend on the 'seed,' the starting number. The seed governs all numbers that come after it. So, even apart from the problem of cycles, the numbers would not be truly random.

Von Neumann would not have been the outstanding mathematician that he was, if he had not been aware of the problems. "Anyone who considers arithmetical methods of producing random digits is, of course, in a state of sin," he famously pronounced. "For…there is no such thing as a random number – there are only methods to produce random numbers, and a strict arithmetic procedure of course is not such a method."

So, as a true generator of random numbers, the middle-square method is out. Worse still: so are all sequences of numbers produced by a computer. By definition, they cannot be random. After all, a computer does exactly as it is told, and only what it is told. 'Pick a random number' is too vague an instruction; the computer cannot follow it. To illustrate this, I now diverge a little to mathematics, in particular to set theory, to invoke the Axiom of Choice.[1]

Lord Chichester has a big collection of shoes and must decide which pair to wear to the Queen's garden party.[2] To make up his mind, he asks his butler to bring exactly one shoe from each of the pairs. The butler has no problem with this request: he simply brings the left shoe of each pair for the lord's

---

[1] What follows is partly based on a chapter in my book *Perplexing Paradoxes*, Columbia University Press, 2024.

[2] This example, though not in these words, goes back to the philosopher Bertrand Russell.

appraisal. Now for suitable socks. Lord Chichester asks the butler to bring one sock from each of the pairs of his sock collection. This time, the butler has a problem. Which sock should he choose? Within each pair, the two socks are identical.

The butler's dilemma points to a deep mathematical question: can choices be made automatically? Some mathematicians believe they can, even if the number of pairs of socks is infinite, others believe they cannot. As disputes among mathematicians go, this was (and is) a particularly fierce one.

It was the mathematician Ernst Zermelo (1871–1953) who provided an answer of sorts. Zermelo investigated a conjecture by Georg Cantor (1845–1918), the founder of set theory, which states that every set of objects can be well ordered. In simple terms, his "Well Ordering Principle" claimed that for any set there exists an ordering, such that a smallest element within that set can be identified. In a bit of a round-about reasoning, this provides the loophole for the butler, but mainly for doubtful mathematicians.

So Zermelo's way out for conflicted mathematicians was simply to assume that choices can always be made. Even better, he stipulated it as an axiom: "For any set X of nonempty sets [i.e., for all the pairs of socks] there exists a choice function f defined on X," that is, some sort of intuition tells the butler which item from the pair to choose. The axiom states that a choice can always be made, without specifying how. This did not end the controversy. To this day, there are two groups of mathematicans: those who subscribe to the Axiom of Choice, and those who do not.

Electronic computers belong to the second group. When instructed to pick a bit, zero or one, they remain motionless like Buridan's ass, unable to make a choice between the left and the right bale of hay. While the poor animal died of hunger, the computers' fate is less dramatic: they simply stop running.

&

The *raison d'être* of computer algorithms is to follow instructions to the dot. Once that is realized, it is clear that computers will never, ever be able to produce random numbers. Hence, the question arose whether and how computer algorithms can be employed to create the raw material needed to do polling, perform simulations, play games, estimate the volumes of high-dimensional bodies, investigate the behavior of neutrons.

The solution was as ingenious, as it was shifty: "If you can't make it, fake it." Simply generate numbers that appear to be random; they will make do for all practical purposes. Well-designed fakes can imitate random numbers;

pollsters, researchers, gamers will be none the wiser.[3] The task is to generate sequences of numbers that satisfy *the three Us*, the holy trinity of randomness: they are *u*npredictable, *u*niformly distributed, and *u*ncorrelated (i.e., independent of each other).

But 'fake' has a nasty ring to it and since branding is everything, a new class of numbers was called to life: not fake, but 'pseudo.' Pseudo-random numbers aren't really random but they seem to be. While they satisfy the first two *Us*, they only pretend to fulfill the third; they are not uncorrelated.

We will have much to say about how pseudo-random numbers are generated but before we do that, let us call attention to the surprising fact that an immense pool of random numbers and digits are close at hand... but cannot be used.

≈

We begin by providing a classification of real numbers, that is, of the numbers which can be expressed in decimal form and be represented on the number lines. They are nested within each other like Russian dolls. At the simplest level, there are the integers (1, 2, 3, ...); then comes the set of rational numbers (1/2, 1/3, 3/7, ...) which contains the integers. Next there are the constructible numbers (that can be constructed with ruler and compass like, e.g., $\sqrt{2}$); they contain the rationals. Then there are the algebraic numbers (solutions to algebraic equations, e.g., $x^5 - 3x + 1 = 0$) which contain the constructibles.

Beyond the algebraic numbers are the transcendentals which cannot be computed by algebraic means. They, in turn, can be classified into computable numbers which can be approximated to any desired precision by an algorithm (e.g., π, e, $\ln(2)$, $\sqrt{2}$, ...) and non-computable numbers which are...well, nobody really knows, except to say that they cannot be computed. So far, so good.

In 1909, the French mathematician Émile Borel (1871–1956) – also a parliamentarian, Minister of the French Navy, and member of the *résistance* during World War II – defined a new brand of numbers: 'normal numbers.'[4] Written in decimal form, each of the ten digits (0,...,9) of a normal number occurs 10% of the time, each combination of two digits (00–99) occurs 1% of the time, each combination of three digits (000–999) occurs one-tenth

---

[3]   Recall that random numbers are required for simulations. A wisecracker could say that with well-designed fakes one can simulate simulations.

[4]   We have met Émile Borel in Chapter 4, as the first modern mathematician to recognize randomness for what it is.

of a percent of the time, and so on. In other words, the digits of normal numbers are uniformly distributed. The set of normal numbers overlaps partly with the constructibles, the algebraics, and the computables; it does not overlap with the integers and the rationals.

Borel did not just introduce the new brand of numbers, he proved an important theorem: *the vast majority of real numbers are normal.*[5] In fact, if you throw a dart at the numbers line, you are bound to hit a normal number. True, there are infinitely many integers and rational and constructible and computable and algebraic numbers, but as numerous as they are, they represent just a blip in the ocean of the real numbers. The probability of your dart landing on a normal number is *one*, the probability of hitting a non-normal number is *zero*. (Somewhat counter-intuitively, this does not mean that it is impossible to hit an integer, a rational number, a constructible or algebraic number; it just means that the probability is zero.)

Now recall that the digits of a normal number are, by definition, uniformly distributed. This is exactly one requirement of a sequence of random digits. Therefore, an obvious suggestion for the generation of random numbers would be to take one of the multitudes of normal numbers and use its decimal expansion as a source of random numbers.

But there's a problem!

As ubiquitous as normal numbers are, we hardly know any of them. It is suspected that many household numbers like $\pi = 3.14159\ldots$, Euler's number $e = 2.71828\ldots$, the natural logarithm of two, $\ln(2) = 0.69314\ldots$, or the square root of 2, $\sqrt{2} = 1.41421\ldots$ are normal. But so far, nobody has been able to prove that. Hence, using their decimal expansions as sequences of random digits remains suspect.

But even if it could be proved that such household numbers are, in fact, normal, there is an even more serious argument against using them as a source of true randomness. There exist algorithms that compute the next digit of the decimal expansions of π, e, $\ln(2)$, $\sqrt{2}$, based on the digits that have been computed so far; hence, the digits are not independent of each other. And therefore, they are not random. At best – if a household number's normality could one day be proven – its digits would at best – if the user is not aware of the generating algorithm – represent a pseudo-random sequence.

---

[5] Almost all real numbers are absolutely normal, in the sense that the numbers that are not absolutely normal form a set of Lebesgue measure zero.

With household numbers out, what do we have? Since Borel showed that the vast majority of numbers is normal, we need to look for a normal number that is not computable. There are at least some that are known, namely the so-called Chaitin constants: they are non-computable normal numbers. But their very non-computability, a requirement for randomness, makes them unavailable for our purposes: we cannot know their digits. Though the Chaitin constants are normal, and their digits random, they are inaccessible. It's a Catch-22 situation: on the one hand, random digits must be uncomputable; on the other hand, since they are uncomputable we cannot know them.

✍

In 1950, at the height of the McCarthy Era's pursuit of suspected communists, the mathematician D. H. Lehmer was fired from his position at Berkeley because he refused to sign the oath of loyalty that the University of California had demanded.[6] This did not stop him from suggesting a new method to generate pseudo-random numbers. He had presented the general idea of an algorithmic random number generator at a "Symposium on Large Scale Digital Calculating Machinery" at Harvard University in September 1949. His landmark contribution was published in 1951 and became the starting point for many further developments of random number generators.

To describe his method – which would become the universal basis for the generation of pseudo-random numbers and, just as importantly, for the encryption of communication, we recall *modular arithmetic*, a special field of mathematics that we already discussed in the context of cryptography.

To illustrate, picture the face of a watch. Let us say it is 11 o'clock and you want to take a 3-hour nap. What time will you wake up? (We'll disregard AM and PM here.) You add 3 hours to 11 o'clock and the answer is 2 o'clock. In mathematical notation, this *modular* operation is expressed as $11 + 3 \bmod(12) = 2$.

Whenever the answer to an arithmetic operation reaches beyond 12 o'clock, the count starts over again: $11 + 38 \bmod(12) = 1$, $6 - 8 \bmod(12) = 10$, $7 \times 3 \bmod(12) = 9$, $5^3 \bmod(12) = 5$. Of course, numbers other than 12 can be used for the base of the modular operation. In fact, usually very, very large bases are used, even several hundred digits long. The important thing

---

[6] A year later, the Supreme Court declared the oath unconstitutional, and Lehmer returned to Berkeley.

is that after the base has been deducted as often as possible, what remains is the result. For example,

$$17 \times 35 \bmod (86) = 595 \bmod (86) = 79$$

Instead of delving into the intricacies of Lehmer's original algorithm we go straight to a more advanced version, the so-called *Linear Congruential Generator* (LCG). Starting with a seed, let's call it X_0 – Lehmer chose the seed at random from a wastepaper basket full of punched cards – one picks a modulus m, a multiplier a, and an increment c, and calculates the sequence of pseudo-random numbers as follows:

$$X_{k+1} = (aX_k + c) \bmod (m)$$

The most important question now is whether the LCG avoids the pitfall of von Neumann's middle-square method, namely that the sequences quickly degenerate into cycles. Unfortunately, cycles are not avoided but with a judicious choices of parameters m, a, c, and X_0, the period can be very long. Another advantage of the LCG is that techniques from number theory actually allow the determination of the periods' lengths. Hence the sequence does not veer into a cycle suddenly and unexpectedly.

But accidents can and do happen. In the 1960s, a popular random number generator was RANDU. Developed by IBM, it became infamous as being one of the most ill-conceived random number generators ever designed. For ease and speed of computation,[7] the designers chose $c = 0$, $a = 2^{16} + 3$, and $m = 2^{31}$:

$$X_{n+1} = (65,539 X_n) \bmod (2,147,483,648)$$

RANDU was put to use in the early 1960s but in 1965 users were in for an unpleasant surprise. Someone (it is no longer known who) discovered that though the numbers were uniformly distributed, whenever three consecutive entries of the sequence were used to plot points in three-dimensional space, the points did not fill the space randomly as they should have. Shockingly, they all fell onto 1 of 15 parallel planes. As random numbers, RANDU's output was useless. Looking back three decades later, Donald Knuth of Stanford University, wrote in the third edition of *The Art of Computer Programming*: "When this chapter was first written in the late

---

[7] Since the length, in bits, of the IBM processor's working registers was 32.

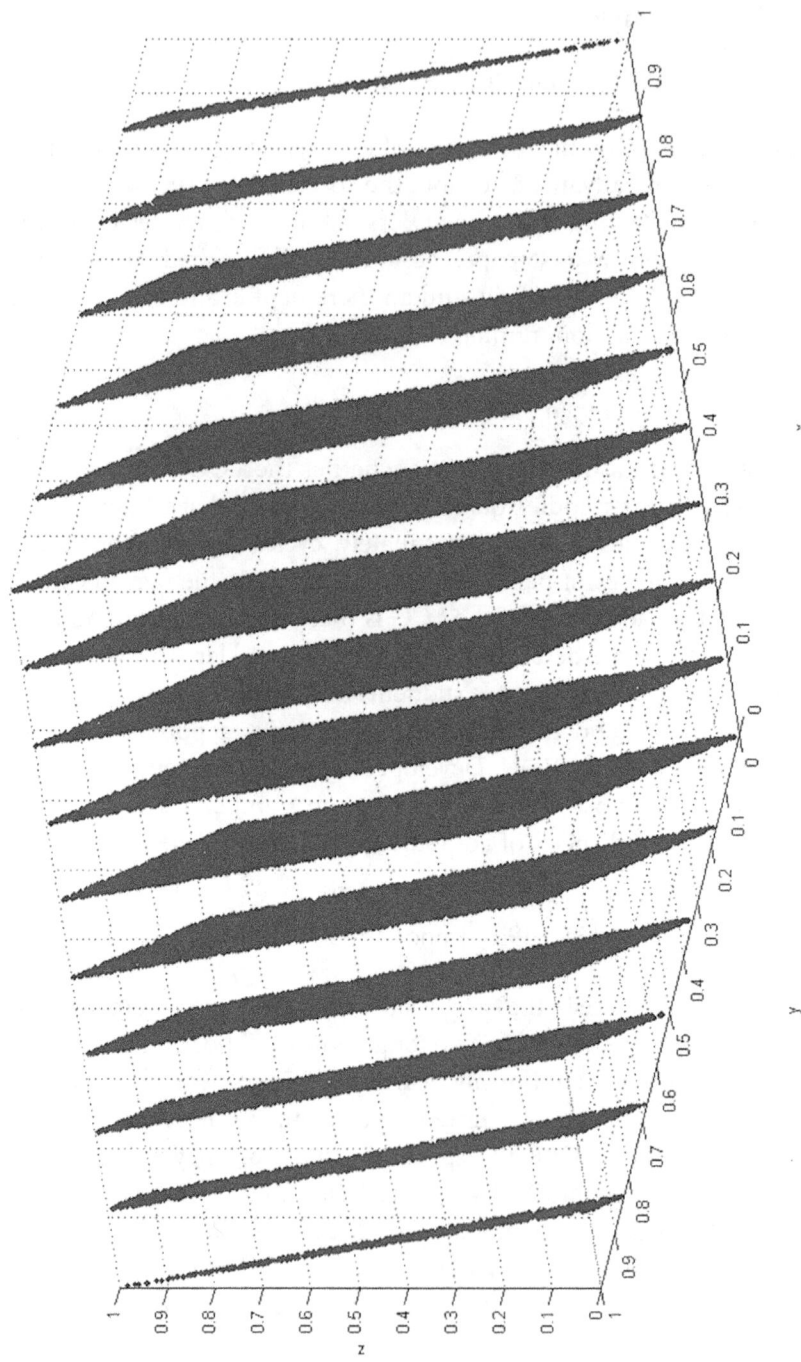

FIGURE 17.1   IBM's random number generator RANDU.
(From https://commons.wikimedia.org/wiki/File:Randu.png).

1960s, a truly horrible random number generator called RANDU was commonly used on most of the world's computers."

The reason for this stark deficiency was discovered by George Marsaglia, a computer scientist at Boeing Scientific Research Laboratories: n-tuples that are produced using successive values drawn from an LCG with modulus m lie in fewer than $(n!m)^{1/n}$ parallel $(n-1)$-dimensional hyperplanes in n-space. The unlucky combination of RANDU's parameters m, a, c was such that the supposedly 'random' numbers fell on 15 discrete planes in 3-space.

The fiasco went to show that the choice of parameters for the LCG (m; a; c) is crucial to produce at least a semblance of randomality. To mitigate the fallout, a proper cottage industry arose around them. Based either on plausibility arguments or on theoretical underpinnings, the Sinclair computer ZX81 used ($2^{16}+1$; 75; 74), C/C++ opted for (2^{32}; 22,695,477; 1), the influential book *Numerical Recipees* proposed (2^{32}; 1,664,535; 1,013,904,223), *Turbo Pascal* recommended (2^{32}; 134,775,813; 1), *Microsoft Visual Basic* suggested (2^{24}; 2,147,483,629; 2,147,483,587), Donald Knuth advocated for (2^{64}; 6,364,136,223,846,793,005; 1,442,695,040,888,963,407) ...

෴

Though the LCG can be useful with a shrewd choice of (m; a; c), Marsaglia's paper actually sounded the death-knell for the LCG. So, the search was on for more sophisticated techniques. One idea that arose was to modify the LCG by using not one but *two* of the preceding entries to compute the new pseudo-random number. This method of generating pseudo-random number is related to what is known as the *Fibonacci sequence*.

What is this sequence? Discovered by the Italian mathematician Leonardo Bonacci (1170–1250), it was a model of how rabbits procreate. A pair of newborn rabbits is released into a field. At age one month, the pair mates and at the end of the second month they produce another pair of rabbits. At the end of the third month, they produce another pair and continue to do so every month. At the same time, the first offspring mate, and, a month later and in all succeeding months, produce a new pair. And, of course, all offspring do the same. (One shortcoming of Fibonacci's model was that the rabbits never die.)

So, the total number of rabbit pairs in the nth month is the number of pairs alive in the previous month (i.e., F_{n-1}) plus the number of new baby rabbit pairs, F_{n-2}. Mathematically, the number of rabbit pairs is expressed as

$$F_n = F_{n-1} + F_{n-2}$$

Starting with $F_0 = 0$ and $F_1 = 1$, one obtains the Fibonacci sequence 0, 1, 1, 2, 3, 5, 8, 13, 21, 34, 55, 89, 144... .

The idea arose to adapt the Fibonacci sequence for the generation of pseudo-random numbers; divide them by 10 and use the remainders as the random numbers: 0, 1, 1, 2, 3, 5, 8, 3, 1, 4, 5, 9, 4, ... Nice try but no cigar! Like von Neumann's middle-square method, this sequence devolves into a cycle too.[8]

Not to be discouraged, scholars realized that instead of utilizing the two entries immediately preceding the sequence's new pseudo-random number, other lags could be employed, thus supposedly leading to 'good' pseudo-random behavior; for example:

$$F_n = \left(F_{n-24} + F_{n-55}\right) \mod (m)$$

with m even, or

$$F_n = \left(F_{n-9,739} + F_{n-23,209}\right) \mod (2)$$

But why use addition to combine the two lagged numbers? One might as well subtract them from each other, or multiply them. And why limit oneself to just two lags? And why not add constants? Another cottage industry opened where variants of LCG and Fibonacci sequence were offered with ever more exotic names – *add-with-carry, subtract-with-borrow, multiply with carry, complementary multiply with carry, linear-feedback shift register* ... – each with its own advantages, disadvantages, idiosyncrasies.

Pseudo-random number generators where each new number depends on two or more previously generated numbers, not just the immediately preceding one, are called 'shift register algorithms' because, as each new number is generated, the previous numbers are shifted by one place.

The RANDU debacle was a failure of a highly organized and well-managed corporation. Maybe a crowd sourced random number generator can do

---

[8] Since there are at most m^2 combinations of pairs $F_{n-1}\mod(m)$, $F_{n-2}\mod(m)$, and each recurrence of a pair results in a recurrence of all following pairs, the sequence can have at most m^2 entries before it repeats.

better? That is why open-source software was developed: to take software development out of the hands of corporations and let a community of developers, academics, and hackers do their thing. The operating system Debian was launched in 1993 as open source. Over the years, it has become one of the most popular and successful operating systems. As of this writing, a dozen versions have been developed, it is installed in millions of individual computer systems, about 1,000 developers all over the world are actively involved in the system's ongoing improvement.

Unfortunately, a serious fiasco occurred in 2006. Developers were busy scanning the latest version of Debian for possible bugs. To do this, they used the analysis tool 'Valgrind.' As it happened, Valgrind hit upon some suspicious code and sent a warning message to the developers which indicated that at a certain point, Debian called an 'uninitialized variable' in order to seed its random number generator. Now, this is a potentially serious bug since, in general, a variable must be set to a certain value before it is called. If, for example, the variable k is needed by the program at some point, it must have been initialized beforehand, say, by setting $k = 0$. If k is called without first specifying what number it is supposed to be, the program will just use whatever number happens to be sitting in its memory location.

The developers came to the conclusion that the source of Valgrind's 'uninitialized variable' error messages were two specific lines of code. They sent messages to colleagues and to higher-ups in the open-source community. "What do you people think about removing those 2 lines of code?" The answer they got was "If it helps with debugging, I'm in favor of removing them." The developers took that as an authorization to delete the two lines and did so forthwith.

What nobody realized at the time was that the call of an uninitialized variable by the random number generator was not a bug but a feature! The variable had been left uninitialized on purpose so that the program would use whatever *random number* just happened to be sitting in the memory location at that time as its seed. It was a particularly clever way of introducing additional randomness into the random number generator.

But the idea was too clever by half. Since the original coders had not bothered to explain the reasoning behind their decision to leave the variables unintialized – there were no comments added to the two lines – the later developers had no clue. So, without further ado, they removed the two offending lines by 'commenting them out' (i.e., by setting them in quotes which ensured that they would be ignored by the program).

Without anybody realizing, this created an immense problem. Deprived of the randomness introduced by the 'uninitialied variable' trick, Debian's random number generator could produce only 32,767 distinct seeds, a laughably small amount.

For 20 months, whenever unsuspecting users of the Debian operating system required a random seed, 1 of only 32,767 numbers was generated. Simulation algorithms utilized the same numbers over and over again. For encryption, Debian's random number generator was completely useless. Nevertheless, for 20 months nobody noticed that something was amiss.

It was only in 2008, that Luciano Bello, an Argentinian package maintainer, a sort of higher officer in the Debian world, called attention to the problem. Within days, the offending quotation marks were removed and the two notorious lines added back in. It is not known whether malevolent hackers had managed to exploit compromised users through brute force attacks during that time.

It is surprising that even after the well-reported Debian fiasco, several years later a similar event occurred. This time the bogeyman was Google. In August 2013, 55 bitcoins (worth at this writing more than twenty million dollars) were stolen from wallets maintained on Android phones. Note that bitcoin transaction must be signed by the owner's private key. In addition, every transaction must carry a number, randomly generated by the owner. If the private key is used together with the same random number for two or more transactions, hackers can compute the key.

A few days after the mishap, a post to Google's Android Developers Blog followed:

> We have now determined that applications which use the Java Cryptography Architecture (JCA) for key generation, signing, or random number generation may not receive cryptographically strong values on Android devices due to improper initialization of the underlying PRNG. Applications that directly invoke the system-provided OpenSSLPRNG without explicit initialization on Androidare are also affected.

In plainspeak, Google informed developers that the Android pseudo-random number generator was supposed to get a 'random' seed from a protected system root file. However, the original programmers had not mentioned this and left it to the user to pick a seed. Of course, most users remained unaware that they needed to do anything, with the result that usually no seed was picked. The result was that bitcoin owners inadvertently applied the same default 'random number' over again when signing their transactions, which allowed criminals to compute their private keys.

Researchers from the cybersecurity firm Symantec explained that "attackers scanned the transaction block chain looking for these particular transactions to retrieve the private key and transfer funds from the Bitcoin wallet without the owner's consent." Shortly thereafter, Google recommended that developers update all apps that use JCA to "explicitly initialize the PRNG with entropy from /dev/urandom or /dev/random."

৵

Pseudo-random number generators devised by corporations can go wrong. Pseudo-random number generators devised by the public can go wrong too. How about letting a corporation design the software and then let the public check it for soundness? Here's an example.

In the mid-1990s, *Netscape Communications Corporation*, originally *Mosaic Communications Corporation*, was one of the very first companies to provide an internet browser, *Mosaic Netscape*. In order to facilitate the emerging field of internet transactions, Netscape developed a cryptographic protocol that relied, of course, on a generator of pseudo-random numbers. Intrigued and somewhat suspicious, two PhD students at Berkeley, Ian Goldberg and David Wagner, decided to analyze the generator in detail. Since Netscape did not make the source code publicly available, they had to resort to the tedious task of reverse-engineering it by manually decompiling the executable program.

To their surprise, they discovered that the seed for Netscape's pseudo-random number generator relied on just three quantities: the time of day (seconds and microseconds), the process ID, and the parent process ID. Now, network analysis tools record the time at which they see each packet. Using the output from such a program, the attacker can guess the time of day on the system running the Netscape browser to within a second. The parent process ID is often just 1, and process IDs are not considered secret information by most applications, so some programs will leak information about them. All that is left unknown are the microseconds of the time. But that just leaves one million possibilities for the seed. An adversary could predict the exact value through a brute force attack and the exact seed would lie open.

The two doctoral students ended their paper by voicing concern about private companies hiding information about their security modules and shunning public review. "Without peer review and intense outside scrutiny of a company's software at the source-code level, there is simply no way consumers can know where there will be future security problems."

Without their sleuthing, the deficient pseudo-random number generator might have remained in use for years. Soon after they announced their successful attack on Netscape's pseudo-random number generator, the company released a new version of the browser which used more randomness in producing the encryption keys. The moral of the story according to Goldberg and Wagner is that software companies must be encouraged to embrace in-depth public scrutiny of their security software.

As of this writing, one of the most advanced and most widely used pseudo-random number generator is the *Mersenne-Twister*, developed by the Japanese mathematicians Makoto Matsumoto and Takuhi Nishimura from Keio University in Tokyo in 1997.

The first part of its name derives from the fact that the period length is a Mersenne Prime (a prime number of the form $2^p - 1$, where p is prime); the second part from the fact that the algorithm is a 'twisted generalised feedback shift register': the Mersenne-Twister generates a new pseudo-random number in the sequence by splitting the bits of two previous entries, 'twisting' them by concatenating the upper bits of one with the lower bits of the other, multiplying the resulting vector with a certain matrix, and combining this result with the bits of another preceding entry.

With proper parameter values, the Mersenne-Twister passes nearly all statistical tests for randomness and has an immensely long period of $2^{19,937} - 1$ (approximately $4.3 \times 10^{6,000}$). Compare this with the number of atoms in the observable universe which is 'only' about 10^{80}. It is not surprising, therefore, that the Mersenne-Twister has become the default pseudo-random number generator for many programming languages and software packages, like Microsoft Excel, C++, Python, Ruby, *Mathematica, Maple, Matlab, …*[9]

A paper in the *Journal of Cyber Security and Mobility* listed ten booby traps of which designers and users of pseudo-random number generators should beware: poor commenting, lack of proper instruction, overly clever

---

[9]   More recently, it failed some tests that were specifically designed to test pseudo-RNG of the form of the Mersenne-Twister. https://arxiv.org/abs/1910.06437.

coding, uncritical use of automated software analysis tools, overwhelming error messages, repairs by non-experts, ambiguous communication, poorly distributed and overly technical announcements, premature posting of patches, and, finally, user community not taking cybersecurity seriously enough or perhaps not having the resources to deal with critical issues.

I close this chapter with a word of warning from the *International Encyclopedia of Statistical Science*: "Do not trust blindly the software vendors. Check the default RNG of your favorite software and be ready to replace it if needed."

CHAPTER 18

Self-Avoiding Random Walks

Monte Carlo Gone Wrong

Similarly to the forging of art or the counterfeiting of coins, the faking of random numbers carries significant risks and can lead to serious consequences. In critical applications, hidden flaws in pseudo-random number generators can spark chaos, compromise security, undermine reliability, and jeopardize accuracy. The hazards go beyond the use of low-quality software which leaves vulnerabilities for hackers to exploit or which botch simulations of high-stakes scenarios. Even seemingly high-quality pseudo-random number generators can go awry when conditions are right...or, rather, when they are wrong. The divide between the usage of truly random numbers and their pseudo relatives is quite perilous. In this chapter, we'll explore some pitfalls and some methods to avoid them.

The embarrassing mishap that occurred in the mid-1960s with IBM's widely used pseudo-random number generator RANDU was not the only fiasco that has come to light. Scientists subsequently improved the algorithms, mainly by replacing linear congruential generators (LCGs) with shift-register algorithms (SRAs), which generate new numbers based on several predecessors that may lie quite far back in the sequence. Though this did help, it was not the end of the story.

DOI: 10.1201/9781003641520-24

In late 1992, a three-page paper appeared in *Physical Review Letters,* one of the most influential journals in physics, which made readers sit up. The title was "Monte Carlo Simulations: Hidden Errors from 'Good' Random Number Generators." The authors, Alan Ferrenberg and David Landau from the University of Georgia, and Joanna Wong of IBM, had been investigating the so-called *Ising-model,* a mathematical description of the magnetization of ferromagnets.

Atoms can have a 'spin' either up or down. If the spins of most atoms point in one direction, the ferromagnet is magnetized; if about half point up and the rest point down, it is not. One question that kept physicists occupied was how different temperatures influence the atoms' spins.

Developed in the 1920s by the PhD student Ernst Ising in Germany, the eponymous model describes how the spins of the atoms align with the spins of the neighboring atoms at various temperatures. Since systems always seek to reach the lowest energy level and a ferromagnet's energy is lower if the spins agree, the spins will tend to point in the same direction if the ferromagnet is placed in a magnetic field, either up or down. What happens when the external magnetic field is turned off? We would expect the system to return gradually to an unmagnetized state. And this is what it does…at high temperatures. At temperatures below a critical level, however, the spins remain stuck in the aligned direction, and the ferromagnet remains magnetized. The temperature at which ferromagnets become magnetized is called a phase transition, the point when a system suddenly changes its property. Other examples of phase transitions are ice melting into water at 0°C, boiling water converting into gas at 100°C, smooth traffic turning into a bottleneck at a certain density, graphite turning into diamond at a certain pressure.

This was the setting for the work of the three co-authors when they discovered something strange. The subject would become known as *Ferrenberg Affair.*

Ernst Ising had investigated atoms in one dimension, arranged along a line. Each atom is connected to its neighbors by a bond, and Ising studied the interactions of their spins. He came to the conclusion that even as temperature changes, no phase transition would occur in the one-dimensional case. Subsequently, without any evidence, he surmised that this would also hold in two dimensions, across a flat grid. (Since ferromagnets are three-dimensional objects, the absence of phase transitions in one- and two-dimensional models did not contradict the phenomenon of phase transitions in real life.)

Ising left research and worked first in the patent office of the electrical equipment company AEG, and then as a high school teacher. But as Adolf Hitler rose to power, he was fired by the Nazis because of his Jewish ancestry; he and his wife survived the war as menial laborers in Luxembourg and in 1947 they left for the United States. Only two years after his emigration was Ising made aware that the work of his student days had become a subject of intense interest among solid-state physicists. But he was no longer interested. Though he became professor of physics at Bradley University in Illinois, he never published any more academic papers.

∾

It turned out that Ising's hunch that no phase transition takes place in the two-dimensional lattice, based on his model of the one-dimensional lattice, was wrong. The Norwegian physical chemist Lars Onsager solved the problem analytically in 1944. (Onsager was awarded the Nobel Prize in Chemistry in 1968.)

With phase transition in the two-dimensional model solved, the race was on to find an analytical solution to the three-dimensional problem. But in spite of much effort by many scientists, there was no progress. Many people nowadays believe that the three-dimensional model may actually be unsolvable and that the best that can be done is to simulate it.

And this is how *Ferrenberg Affair* began. Maybe the designation 'affair' is a bit over the top but in the staid world of science the news certainly was a shocker.

Until the late 1980s, the habitual method of studying simulations of the Ising systems was to select a site at random and to determine its and its neighbors' spins. Next, the energy differential was calculated that would occur if the site's spin were flipped. (If its spin would align with a majority of the neighboring spins, the energy would be reduced, and vice versa.) The probability that a spin actually flips depends on the energy differential and the prevailing temperature. Then came the crucial step in the simulation: if a random number between zero and one was less than the calculated probability, the spin was flipped, otherwise not. The procedure was repeated with millions of random sites, until the system reached an equilibrium. That is how Ising systems used to be simulated.

Unfortunately, as the simulated temperature of the system descends towards a critical level, the algorithm begins to require vastly more Monte Carlo steps; this slowed down the simulation, rendering it all but impractical. To overcome the problem, the German physicist Ulli Wolff from Kiel University implemented a new method of modelling three-dimensional Ising systems. Spins would not be flipped individually, one by one, but clusters of sites would be established, depending randomly on a probability determined by the energy and temperature. Then the entire cluster would be flipped at once. Since Wolff's algorithm produced the same results as the one-by-one flipping method, but without the slowing down, it was considered acceptable.

Back to Ferrenberg and his colleagues, before attempting to apply Wolff's algorithm to the knotty problem of three-dimensional Ising systems, they subjected it to a test in two dimensions. After all, in flat space, the method can be validated by comparing simulation results with the precise analytical results that Onsager's computations provided.

To generate random numbers, the scientists used a two-shift-register algorithm and the bit-wise XOR operation (The operation 'exclusive or,' denoted by $\oplus$, compares the bits of the two operands. If they agree – both are zero or one – the resulting bit is zero, otherwise one.) Then they ran simulations and compared them with Onsager's analytical computations. They should have been identical.

They weren't!

True, the differences in absolute values were minute, but they were statistically very significant because with millions of steps, the Monte Carlo simulations gave very sharp results. Sharp but incorrect. In two of the cases, the simulation results were 42 and 107 standard deviations removed from the correct analytical results.

The three scientists were aghast and the question now was what had gone wrong. The cluster-flipping algorithm itself had been in use for several years and had passed all tests. Furthermore, when Ferrenberg ran the Wolff algorithm with random number generators other than the two-shift-register algorithm, the results were acceptable; they fell within one standard deviation from the analytical results. So, suspicion fell on the two-shift random number generator. But that alone could not be the answer either, since it had proved itself with many, many other simulations.

It turned out that the cause for the problem was neither the cluster-flipping algorithm by itself nor the random number generator by itself; it was the *interaction* between the two. It was the use of that specific algorithm with that specific pseudo-random number generator that triggered the faulty results. This was very surprising indeed; nobody had previously considered the possibility that a 'good' algorithm and a 'good' random number generator would become 'bad' when used in conjunction.

Ferrenberg et al. explained the faulty results with subtle correlations in the random number sequence which affect the Wolff algorithm in a special way. Sequences of random numbers may appear, in which the leftmost bits are zero. Since the XOR operator renders the combination of two zeros to zero, these bits may remain zero in subsequently generated numbers which, in turn, may lead to a bias in the size of the Wolff clusters. Additional evidence of some subtle serial correlation in the supposedly random numbers was given by the fact that when the simulation used only every fifth generated number, the bias disappeared.

The authors concluded with some pertinent advice for Monte Carlo simulations: even if a random number generator passes all traditional checks, "a specific algorithm must be tested together with the random number generator being used."

It did not take long for others to take up the cudgel. Peter Grassberger from Wuppertral University in Germany was interested in random walks, that is, walks that drunken sailors would take after leaving a bar if they randomly make steps forward or backward. He simulated random walks with random bits: if it is zero, the sailor takes a step forward, if it is one, he totters backwards. One question is, where a drunkard will find himself after N steps. Another question is whether a certain point in higher-dimensional space can be reached if parts of the path are blocked. The answer to the first question is that if he staggers along a line, he will find himself on average at $\sqrt{N}$ steps in either of the two directions from where he started.

The random walks that Grassberger studied were a bit more complicated. First, they took place not along a line, but on a three-dimensional lattice. (Three-dimensional walks describe, e.g., how water percolates through coffee grinds to produce an *espresso*.) Second, the drunkard never visited the same lattice point twice. Such random walks are called self-avoiding walks. To simulate random walks in three dimensions, random numbers provide the probabilities of the walker taking a step into any of the six directions (up, down, left, right, front, back).

Grassberger had been using pseudo-random number generators for years to study the number of self-avoiding walks that exist in a three-dimensional lattice. Now, alerted to potential problems by *Ferrenberg Affair*, he decided to take a step back (no pun intended) and reconsider his previous work.

For up to 23 steps ($N = 23$), mathematicians had enumerated all possible self-avoiding walks on a three-dimensional lattice; hence their total numbers were known exactly. (For example, for 3 steps there are 150 possible self-avoiding walks, for 4 steps there are 726, for $N = 23$ there are 5,245,988,215,191,414.) The purpose of Grassberger's Monte Carlo simulations was to verify whether the number of simulated walks increases at the same rate as the enumerated walks. In principle, as the number of steps increases, samples created by Monte Carlo simulations should grow at the same rate as the enumerated self-avoiding walks. Hence, normalized with appropriate factors, the number of simulated walks, $n(N)$, divided by the actual number of walks, $C(N)$, should be about 1.0.

Using seven different pseudo-random generators, Grassberger simulated millions of self-avoiding walks and plotted the normalized ratios $n(N)/C(N)$, for N, the number of steps, between 1 and 23. What he saw was both surprising and shocking.

With a conventional LCG and with judiciously chosen parameters,

$$X_n = \left(69{,}069 X_{n-1} + 1\right) \bmod \left(2^{23}\right)$$

the ratios were in fact, very close to 1.0 for all step sizes between 1 and 23. So far, so good.

But ever since the notorious RANDU debacle, it was known that LCGs may harbor problems that lurk somewhere and suddenly appear when least expected. Hence, Grassberger decided to give the more sophisticated *Shift-Register Generators* SRGs a try. These generators, that use several preceding entries to compute the new one, had become all the rage. Grassberger used three versions that had proved themselves worthy in other work:

$$X_n = X_{n-5} + X_{n-17} \bmod \left(2^{32}\right),$$

$$X_n = X_{n-24} + X_{n-55} \bmod \left(2^{32}\right),$$

and $X_n = X_{n-103} + X_{n-250} \mathrm{mod}\left(2^{32}\right)$

To his dismay, results were not good at all. For each one of these SRGs, the normalized ratios $n(N)/C(N)$ deviated considerably from the hoped for 1.0. So Grassberger tried another track, namely to use the operator XOR instead of addition to combine the two entries (which, by the way, lay far in the past):

$$X_n = X_{n-103} \oplus X_{n-250} \text{ and}$$

$$X_n = X_{n-1,063} \oplus X_{n-1,279}$$

These simulations did not fare any better; the normalized ratios were again off, deviating from 1.0 by up to 7%. On the other hand, a different SRG with no less than four inputs, which had proved itself in various simulations by other researchers,

$$X_n = X_{n-157} \oplus X_{n-314} \oplus X_{n-471} \oplus X_{n-9,689}$$

allowed the algorithm to simulate walks whose ratios were – surprise, surprise – close to 1.0. Grassberger concluded that a triple correlation between the new random number, X_n, and the two previous entries, X_{n-p} and X_{n-q}, may – but need not – be hidden in the series of numbers produced by SRGs.

It is easy to show that pseudo-random number generators that are based on *Linear Feedback Shift Registers* (LFSR),

$$X_k = X_{k-p} + X_{k-q} \mathrm{mod}(m)$$

must eventually lead into cycles. After all, given the modulus m, there are at most 2^m different sequences that can appear. Why? Because whenever one of those sequences appears, the algorithm must recreate the subsequent numbers, as it did when it first occurred. Furthermore, the sequence is fully determined by the previous p or q generated numbers (whichever is larger).

In the best of cases, the cycle will be as long as the modulus, and if that is sufficiently large, the generated series may be adequate to serve as a source of random numbers. But in 2004, two German physicists, Heiko Bauke and

Stephan Mertens, unexpectedly proved that even if the series is sufficiently long, generators based on LFSR are flawed. They showed, both mathematically and by numerical experiments, that an LFSR generator of binary digits that simulates coin flips (1 being 'heads,' 0 being 'tails'), produces fewer 'tails' than 'heads.'

That is truly surprising. What's special about 'tails? Bauke and Mertens retraced the dearth of tails to a special characteristic of the number zero in arithmetic. To explain, we must make a slight detour to group theory, a special branch of mathematics.

A *group* is defined as a set of numbers together with an operation that combines two numbers such that the result is again contained in the group. Importantly, the set must also contain a neutral element which leaves the result unchanged ($3 + 0 = 3$), and an inverse to every element ($A + (-A) = 0$). Let us look at an example:

The set of real numbers is a group under addition because the sum of two reals is also a real. The number zero is called the neutral element, since adding zero to a real number leaves the result unchanged. And to every real number Q, there exists an inverse, $-Q$, which is also a real number.

Is the set of real numbers a group under multiplication? Well, the product of two real numbers is also a real number. And the neutral element is 1, since multiplying a real number by 1 leaves the result unchanged. But there's a snag: though 0 is also a real number, the inverse, 1 divided by 0, is not a real number; in fact, $1/0$ is not a number at all. So, the set of real numbers becomes a group only *if the number zero is excluded*.

Hence, the existence of the number zero, perfectly good under addition, prevents the set of real numbers from being a group under multiplication. Only when the number zero is removed from the set, do the reals form a multiplicative group.

This is the underlying reason why bits produced by LFSR generators show more heads than tails. An indication of why the problematic nature of the number zero leads to a bias against tails can be gleaned from the following fact. In the starting sequence, any combination of zeros and ones is possible ... with one notable exception: an all-zero starting sequence, (0, 0, 0, ..., 0), would result in nothing but zeros in the entire sequence.

Bauke and Mertens did not only identify the problem but also came up with a workaround. They suggested to use generators that produce not just zeros and ones, but zeros, ones, and twos. Such algorithms are easily implemented with LSFRs with modulus 3. The ones that are generated are considered 'heads,' the twos 'tails,' and the zeros are simply ignored.

❧

Monte Carlo simulations are extremely useful in situations when answers to questions are unknown or computations are impossible. But their use requires trust in the generator of the pseudo-random numbers. Unfortunately, that trust may be misplaced. We have reviewed three examples of Monte Carlo simulations that gave verifiably incorrect results due to imperfections to the use of pseudo-random number generators. These may be only the tip of the iceberg, however, visible solely because researchers specifically looked for glitches. Many more Monte Carlo-based studies may have been, and may still be, seriously compromised due to poor quality random number generators, without the unwitting users ever realizing it.

The matter reminds of a Catch-22 situation. On the one hand, if the simulation results are close to what is expected, preexisting beliefs are simply reaffirmed and the entire exercise may have been superfluous. On the other hand, if simulation results are far from what is expected, one finds oneself in a dilemma: were the expectations wrong or is the simulation faulty?

Two physicists at CERN, the nuclear research facility in Geneva, suggested a due-diligence audit that should be carried out before one makes use of pseudo-random number generators:

1. The period should be much longer than any sequence that will be used in any one [simulation], but a long period is not sufficient to ensure lack of defects.

2. Empirical testing can demonstrate that a RNG has defects (if it fails a test), but passing any number of empirical tests can never prove the absence of defects.

3. Making an algorithm more complicated (in particular, combining two or more methods in the same algorithm) may make a better RNG, but it can also make one much worse than a simpler component method alone if the component methods are not statistically independent.

4. It is better to use a RNG which has been studied, whose defects are known and understood, than one which looks good but whose defects are not understood.

5. There is no general method to determine how good a RNG must be for a particular Monte Carlo application. The best way to ensure that a RNG is good enough for a given application is to use one designed to be good enough for all applications.

Of course, none of these recommendations guarantee true, or close to true, randomness. John von Neumann's quotation, cited in the previous chapter, remains depressingly pertinent: "there is no such thing as a random number – there are only methods to produce random numbers, and a strict arithmetic procedure of course is not such a method."

We end this chapter with slightly more constructive advice given by the computer scientist Donald Knuth of Stanford University: "…random numbers should never be produced by a random method. Some theory should be used."

So be it!

CHAPTER 19

Squeezing Randomness Out of Lemons

Extractors

As we emphasize throughout this book, random numbers are not easy to come by. The only good sources of true random numbers are physical implements (like coins or dice or floating ping-pong balls), physical phenomena (like radioactive decay, thermal noise, number of sunspots), human jitters (like movements with the computer mouse, keyboard typing), economic phenomena (like the movement of the stock market, the prices of commodities). They, as well as random number sequences generated by computer algorithms, may pass all randomness tests and nevertheless have some hidden regularity that has simply not yet been discovered. So, how can we obtain random numbers in a world without perfect randomness? In this chapter, we investigate how a series of random numbers can be generated if all that is at our disposal are one or more series of not quite random numbers.

For example, can a truly random string of bits be generated if we have a coin that does not fall heads or tails equally often. Say the coin produces 60% heads and 40% tails. Can we still use it to create an unbiased string of zeros and ones?

Surprisingly, the answer is yes. Let us denote the probability of throwing heads by q and the probability of throwing tails by $1 - q$. To produce one

DOI: 10.1201/9781003641520-25

random number, we throw the coin twice. The probabilities of getting any combination of two throws are

Heads then heads	$q \times q$	36%
Heads then tails	$q \times (1-q)$	**24%**
Tails then tails	$(1-q) \times (1-q)$	16%
Tails then heads	$(1-q) \times q$	**24%**

The probability of getting heads then tails, and tails then heads, is equal: 24%. Thus, we can extract a true random number sequence from the biased coin throws by entering the bit *one*, whenever the two-throw result is *heads then tails*; the bit *zero*, whenever the two-throw results is *tails then heads*; the two-throw results *head–heads* and *tails–tails* will simply be ignored. The resulting sequence will be 50% ones and 50% zeroes, even though the coin was severely biased.[1]

The creator of this method was none other than John von Neumann in 1951. Though it is only about one quarter as efficient as tossing truly unbiased coins – with q close to 0.5, two double-throws are required, on average, to generate one random bit – the resulting series is truly unbiased. By the way, the coins' actual bias is irrelevant; von Neumann's method works for any value of q.

But what if, for some reason, the current flip depends on the previous flip? Though unlikely for coins, such dependencies, called Markov processes, do appear frequently in nature. For example, the probability of rain today depends to some degree on whether it rained the previous day, and vice versa; but it does not depend, or to a much lower degree, on whether it rained two or three days ago.

It was the celebrated scientist Paul Samuelson who dealt with this problem in 1968, two years before he would receive the Nobel Prize in economics. He suggested two improved versions of what he called the "von Neumann trick." For Version 2.1 he noted that one may use coin flips that are separated from one another by more than one period; this reduces the dependence of the flips on each other significantly. For example, even if coin flips did depend on their predecessors, flip t and flip $t - 10$ would be much less dependent on each other.

---

[1] This is so, notwithstanding the fact that in a sequence of random coin flips, there are more *HTs* in the sequence than there are *HHs* (see Segert, 2024).

(c) ChatGPT

Version 2.2 proceeds in two stages. In the first stage, double-throws are performed again, but only the pairs with a leading H are considered. By starting with a common element (H in our case, but it could equally be T), the Markov dependence of the following elements is broken. By focusing only on pairs with a leading H, a uniform starting condition is created for each new throw. This removes the influence of any previous flips that led to the H, thereby breaking the Markov dependence. Essentially, it ensures that each subsequent throw is only influenced by the fixed starting state and not by any previous throws.

Pairs of the form HH, which appear with probability p_1, are transformed into 1s, pairs HT, with probability p_0, are transformed into 0s. (TH and TT are ignored.) In the second stage, the von Neumann trick is applied to this new series of 0s and 1s: pairs 10, with probability $p_1 p_0$, generate the random bit 1, pairs 01, with the identical probability $p_0 p_1$, generate the random bit 0. (Pairs 00 and 11 are ignored.) Though even less efficient than von Neumann's original trick – Version 2.2 requires eight coin flips, on average, to generate one random bit – the new series consists of truly unbiased and independent random bits.[2]

✎

Chaos theory, which became all the rage in the 1990s, provides another method of extracting random numbers from an imperfect source. The emblem of chaos theory is the sensitivity that a phenomenon exhibits to the initial conditions of the system: due to the non-linearities inherent in the system, the tiniest imprecision in the measurements of the position, speed, weight, acceleration, or any other explanatory variable of the object being studied causes results to diverge substantially. Chaos theory is famously illustrated by the so-called butterfly effect which says that the weather in one location depends so sensitively on the initial conditions, and be it only at, say, the twentieth digit after the decimal point, that a butterfly flapping its wings in Australia may cause a storm in Texas.

The effects of chaos theory can be illustrated in a nutshell by the numbers X_i ($i = 1,2,3,\ldots$) that are generated by the innocent-looking recursive equation (called logistic equation).

$$X_{i+1} = 4X_i \left(1 - X_i\right)$$

Since each entry in the series is determined by its predecessor, we realize right off the bat that this is totally counter to what a random number series demands … except for one detail: the series is totally chaotic. Starting with $X_1 = 0.1$, we get 0.360 as the second entry and then 0.921…, 0.289…, 0.821…, 0.585… . If you did not know how the numbers were generated, you could have been fooled to believing that they were quite random.

---

[2]   Later, Manuel Blum considered sources generated by finite-state Markov chains.

So why not make use of the logistic equation to generate pseudo-random numbers? Well, there's a problem. Though the logistic generates numbers between zero and one, they are not uniformly distributed. Starting again with $X_1 = 0.1$, the first 4,000 entries are distributed as follows:

0 to 0.2	1,166
0.2 to 0.4	576
0.4 to 0.6	524
0.6 to 0.8	532
0.8 to 1.0	1,202

Towards the middle of the range, they are distributed in a fairly uniform manner; but at the low and the high ends they are bunched together. Obviously, this is unacceptable for pseudo-random numbers. But the logistic was too good an opportunity to pass up. Hence, the question arose whether a series of pseudo-random numbers could be extracted from the logistic equation such that its entries would be distributed in an acceptable manner.

And indeed, Stanislas Ulam, the Los Alamos mathematician who pioneered Monte Carlo simulations, together with his colleague P. R. Stein, reported in 1964 on such a method. By transforming the entries of the logistic equation according to

$$y_i = (2/\pi)\sin^{-1}\left(\sqrt{x_i}\right)$$

the resulting y_i are uniformly distributed throughout the interval (0,1):

0 to 0.2	751
0.2 to 0.4	830
0.4 to 0.6	831
0.6 to 0.8	757
0.8 to 1.0	831

For all its faults – the entries are completely predictable – the series extracted from the logistic equation fulfills at least the requirement of uniform distribution and could therefore be mistaken for a pseudo-random sequence.

Von Neumann's trick showed that it is possible to extract perfectly random numbers from a lemon, that is, from an imperfect source. Are there other techniques or tricks? Yes, there are, but there's a require-ment: the series must contain at least some randomness. To wit: if the coin falls tails 100% of the times, there is no randomness at all and nothing can be extracted. Only if the coin has even just a small probability of falling heads does von Neumann's trick allow the extraction of a series of perfectly random numbers. Is this possible with other lemons, that is, with generators that produce strings of random-looking, albeit imperfectly random, numbers?

Before we begin our investigation, we must quantify the amount of randomness that a source contains… or the converse: how much information a source provides. The key idea, introduced by Claude Shannon (remember him, trying to beat casinos at Las Vegas?), is the concept of entropy which is a fancy word for 'disorder.' The higher a string's disorder (i.e., its randomness), the higher the entropy. Before we define the concept, let us define the information content of a coin toss.

A device with two positions, say a coin, can store one bit of information (heads or tails, or zero or one). N such devices (or throws) can store N bits, hence the number of possible states is 2^N, and since $N = \log_2 2^N$, Shannon defined the information content that one gains when the coin falls 'heads,' measured in bits, as

$$I = \log_2\left(1/p_H\right)$$

where p_H is the probability of the coin falling heads.

In what follows, we need a new concept, closely related to Shannon entropy, namely '*min-entropy*,' designated as H_∞. Min-entropy, a measure of unpredictability, refers to the minimum amount of uncertainty or unpredictability in a system. Specifically, it focuses on the most likely or frequent outcome and measures the minimum amount of information associated with that outcome. Essentially, min-entropy quantifies the amount of uncertainty in a system by measuring the minimum amount of entropy present, corresponding to the highest probability event…

Imagine a list of different outcomes, and one of them happens most often. Min-entropy tells us how surprising or unexpected that most frequent outcome is. The higher the min-entropy, the less predictable the most likely outcome.

The next few paragraphs may be skipped by readers not interested in the technical details.

The use of the logarithmic function is quite plausible, since, if $p_H = 100\%$, throwing the coin gives no information whatsoever and, indeed, $I = \log_2(1) = 0$. On the other hand, the largest amount of information is obtained when the coin is fair: if $p_\mathrm{H} = \frac{1}{2}$, the information content of a toss is

$$I = \log_2\left(1/p_H\right) = \log_2\left(2\right) = 1$$

that is, the toss provides one bit of information. The throw of a six-sided die, on the other hand, provides more than two and a half bits:

$$H = 6\frac{1}{6}\log_2\left(\frac{1}{6}\right) = 2.584\ldots\text{bits}$$

which, again, is quite plausible because the throws of fair dice are much more surprising than the throws of fair, and especially of biased coins. The entropy of throws of a loaded die that comes up 1 and 6 with probabilities of 20% each, and 2, 3, 4, and 5 with probabilities of 15% is

$$H = 0.4\log_2\left(0.2\right) + 0.6\log_2\left(0.15\right) = 2.571\ldots$$

Entropy is defined as the average information content of a *sequence* of coin throws, namely the weighted average of the information content of the throws. When a fair coin is thrown N times, one obtains an N-tuple of random zeros and ones. The probability of each possible N-tuple is $p_N = \frac{1}{2}^N$ and the information content of each N-tuple is $I = \log_2 2^N$ which corresponds to N bits.

To illustrate, consider, triples like 110, 101, … On condition that the coins are unbiased, each of the $2^3 = 8$ possible triples appears with the probability $p = \frac{1}{8}$. Hence, the entropy is $\log_2(8) = 3$, and the information content of a triple of coin throws is three bits.

This is the information content of a single N-tuple. Let us advance another step. What is the entropy of a string of N-tuples, or of the process that generates the string? Shannon defined it as the average information content of all possible tuples, weighted by the probabilities p_i of their occurrences:

$$H = -\sum_I p_i \log_2\left(1/p_i\right)$$

In other words, the entropy of a string of tuples is the information that one expects to gain from each tuple. (The minus sign arises to correct for the fact that the logarithm of a number smaller than one is negative.) To illustrate, consider four-tuples, 0000, 0001, 0010,... The entropy of the entire set of $2^4 = 16$ four-tuples is

$$H = \sum_{i=1 \text{ to } 16} \frac{1}{16} \log_2 \left(\frac{1}{1/16} \right) = \frac{16}{16} \log_2 (16) = 4$$

A coin that falls heads with a probability of 90% does not surprise us 90% of the times; it's what we would have expected and the information content is low. Only if it falls tails are we surprised, and the information content is much greater. Hence, on average, the information that the throw of such a coin provides is less than half a bit of information:

$$H = 90\% \text{ of } \log_2 \left(\frac{1}{0.9} \right) + 10\% \text{ of } \log_2 \left(\frac{1}{0.1} \right) = 0.468\ldots \text{bits}$$

and N throws with such a biased coin provide $0.468\ldots \times N$ bits of information.

Back to our question of how much true randomness can be squeezed out of a lemon. For a generating process with N possible outcomes, each of which occurs with probability p_i, *min-entropy*, H_∞, is defined as the entropy of the outcome that carries the least information, that is, the outcome which occurs most frequently and therefore creates the least surprise: [3]

$$H_\infty = -\log_2 \left(\max_i p_i \right)$$

Min-entropy is equal to k if the probability of hitting any specific value is equal to 2^{-k}. It is greater than k if the probability of hitting any specific value is at most 2^{-k}. While a coin that is biased 90% towards heads has Shannon entropy 0.468... (shown previously), it has min-entropy of only $-\log_2 (0.9) = 0.152\ldots$. Therefore,

---

[3] The Min-entropy can also be expressed as $H_\infty = -\min_i \log_2(1/p_i)$.

one requires $^1/_{0.152...}$ occurrences (i.e., 6.578... throws of the biased coin) to get one truly random bit. Note that $0.9^{6.578...} = 0.500$.

Hence, H_∞, the worst-case measure of a sequence's randomness, is the number of random bits that can, at best, be squeezed from a defective pseudo-random source. In order to be able, at least in principle, to squeeze k truly random bits from a sequence that is imperfectly random, this lemon must have min-entropy of at least k.

৵

In the quest to extract high-quality random numbers from low-random sources, three methods have proved themselves: (A) use one low-random source together with some truly random seeds, (B) combine several low-random sources (each of which possesses a certain level of min-entropy), and (C) do away with randomness altogether and utilize computational difficulty instead.

(A) To start, let us see what can be achieved if one has just a single, low-quality random source. Since it has a certain level of min-entropy, it must contain some hidden randomness. But where? And how can we extract it?

Computer scientists have spent decades on efforts to extract truly random bits from the muddle of low- and high-quality bits, jumbled together in the source. Let us say that hidden among the n bits of a source there are k bits that are truly random; the other $n - k$ bits are biased. In order to extract as much of whatever high-quality randomness is hidden in the source, the scientists devised so-called extractor algorithms: first, a small truly random seed is chosen and employed as a catalyst; then the extractor combines the source's n bits with the seed's d bits, churns them around and around, and spits out m bits that are close to uniformly distributed. Hence, they are close to truly random.[4]

How close? This requires another definition. A distribution is said to be ε-close to the uniform distribution if the probability of deviations from the uniform distribution (above or below) does not exceed ε. (The throws of a coin that lands heads with probability 0.52, and tails with 0.48, are said to be 0.04-close to uniform.)

Extractors have five parameters. Apart from the number of bits that the source contains (n, which is given at the outset), there is the min-entropy k; the length of the seed d; the number m of random bits that are

---

[4]   Such extractors are called *seeded* extractors to distinguish them from extractors that do not require a seed (as, e.g., the von Neumann trick).

extracted; ε (the closeness to uniformity). Using a method devised by Paul Erdös,[5] one can prove that for any chosen values of n, k, and ε, there exist extractors that are able to generate a quite respectable number of truly random bits, as long as the seed is sufficiently long.[6]

But one wants to do better than that. One wants to get m to be as close as possible to k, and to minimize the length of the seeds d, while keeping the deviations ε small. Moreover, one also wants the extractor to be computable in an efficient manner.

For several decades, scientists strove to devise ever better seeded extractors and we are now at the stage where the number of extracted random bits, m, is nearly as large as k (i.e., essentially all of the source's min-entropy is extracted), the required seed length d is only of the order $\log_2(n)$, and the error, ε, has been lowered to $1/\log_2(n)$.[7]

But adding a truly random seed to an inferior source may seem a bit like kicking the can down the road. After all, the reason to add a seed as a catalyst was to address the problem that even seemingly random phenomena are not truly random. So, in order to circumvent the need for truly random numbers, one requires truly random seeds. It's a *Catch 22* situation.

Is there another way of squeezing high-quality random digits from low-quality sources, one that does not require conjuring up truly random seeds? There is, and it is surprisingly simple:

(B) Say, we have two independent bit-sequences $A = a_1, a_2, a_3, ..., a_m$, and $B = b_1, b_2, b_3, ..., b_m$. Both of them are not truly random, so neither A nor B on their own can serve as sources of randomness. But on condition that they each possess at least $m/2$ of min-entropy – that is, at least half of the bits are truly random – their bits can be 'blended' (as in a food processor) to provide truly random bits.

It has been proved in the 1980s that the parity of the so-called dot product of sub-sequences of the two independent bit-sequences A and B, each divided into sub-sequences of length n, produces bits with high randomness. Specifically, for each pair of sub-sequences from A and B, the dot

---

[5] This method is called *probabilistic method*. It allows Meto prove the existence of an object that possesses specified properties, without actually constructing the object. In our case, the object is the extractor.

[6] If the seed is at least $d = \log_2(n - k) + 2\log_2(1/\varepsilon) + $ constant, the number of truly random bits is $m = k + d - 2\log_2(1/\varepsilon) - $ constant.

[7] Researchers like Noga Alon, Y. Z. Yao, and others have built on the foundational work of pioneers such as Paul Erdös. Their work includes explicit constructions of robust and efficient extractors that approach the theoretical limits established by Erdős and his contemporaries.

product $C = a_1b_1 + a_2b_2 + a_3b_3 + \ldots + a_nb_n$ is computed, and the parity of C (i.e., $C \bmod 2$) produces a bit with high randomness. Hence, dot products of the two sub-sequences modulo 2 qualify as random zeros and ones. Using no more than the arithmetic operators of addition and multiplication, two independent sources can serve as a good one-bit extractor as long as each sub-sequence contains a proportional amount of min-entropy, ensuring that at least $n/2$ bits are truly random bits for sub-sequences of length n.

The choice of n determines the quality of the new sequence. The longer the sub-sequences, that is, the more bits are involved in the dot product calculation, the better the randomness quality. Shorter sub-sequences might not provide enough mixing of bits, leading to lower quality of randomness. The choice of n depends on the desired trade-off between the number of random bits one wants to generate and the statistical quality of those bits.

Another method of producing truly random bits is to combine several unpredictable though not truly random sources. Say, we have data on sunspots s_i, the stock market m_i, and the weather w_i. For all we know, these data series are only weakly random. But if each source has min-entropy of at least k, the arithmetic operations of addition and multiplication can again be used to extract true randomness from a convolution of these three data series. Then, for example,

$$F\left(s_i, m_i, w_i\right) = s_i \times m_i + w_i$$

is an extractor that produces bits b_i that are not further removed from uniformity (i.e., from true randomness) than 2^{-ek}, for some constant e. These new bits can then be used recursively again and again to generate bits that are even closer to true randomness.

(C) Now to the *non plus ultra* of randomness extractors. Is there anything that can be done if one has no access to randomness at all? If entropy is zero? What if true randomness does not exist, that everything – coin tosses, sunspots, stock market movements… – is biased, correlated, or predetermined?

Surprisingly, something can be done even in that case! One can use computationally hard functions (to be defined further) to create sequences of digits that look random to a randomized algorithm. Now, that sounds suspiciously like pulling a rabbit out of a hat: create something (randomness) out of nothing (no entropy). We must delve further into the matter.

In a paper mentioned previously, Russell Impagliazzo and Avi Wigderson showed that in order to create n pseudo-random bits, seeds with only

$O(\log_2(n))$ truly random bits are required ... if a certain condition is fulfilled. To generate, say, a million pseudo-random bits, the seed would need to be only about 20 digits long. Importantly, however, the required condition is that so-called hard functions exist.

Functions are defined as 'hard' if they are easy to compute but hard (i.e., difficult) to invert. To illustrate, think of multiplication, which is easy, and division, which is based on trial and error and therefore more difficult. On the other hand, if a solution to a hard problem is suspected, it is easy to verify it. To wit, it is hard to find the prime factors of a large semi-prime, but if one of the factors is known, it is easy to verify that it does, in fact, factor the semi-prime.

On a more advanced level, think of multiplications modulo a number: easy to perform, easy to verify, hard to invert. Thus, modular arithmetic is hard. In principle, hard functions can be inverted, but it would usually take an astronomically long, namely exponentially long, time. This is why hard functions are also called *one-way* functions.

Unfortunately, and somewhat surprisingly, it is not known whether hard functions actually exist. True, to date, no algorithm has been found that would invert a hard function in reasonable (i.e., polynomial) time. But maybe, someday, someone, somewhere will develop an efficient algorithm that will be able to invert hard functions?

It is the notorious question of whether $P = NP$, where the class NP (non-deterministic polynomial time algorithms) consists of decision problems (yes/no problems) and where a 'yes' answer can be verified in polynomial time.[8] For example, while it is hard to find the factors of a large integer Q, one can easily verify in polynomial time whether Q is a composite number if given a suspected factor: simply divide Q by the suspected factor; if the division is possible, the answer to the question 'is Q composite?' is 'yes'. Hence, this decision problem is in the class NP.

Whether $P = NP$ is not known. For now, all that is known is that P is a subset of NP; anything that is in P is also in NP. To illustrate, until 2004, the problem of determining whether an integer is prime was only in class NP. With the publication of the landmark paper "PRIMES is in P" the problem moved into P.

If it should turn out that $P = NP$, then hard functions do not exist. However, Impagliazzo and Wigderson assumed, as do most computer

---

[8] The Clay foundation has promised a 1-million-dollar prize to the person who proves or disproves $P = NP$.

scientists, that $P \neq NP$, which is tantamount to saying that hard functions do exist. So let us also assume that $P \neq NP$ and let us follow Impagliazzo and Wigderson's recipe: pick a seed of length $O(\log_2 n)$ and create n 'sub-seeds'; apply each of the sub-seeds to the hard function and create a bit, altogether n bits. To a polynomial-time algorithm, the generated bits look totally random.

You might say that the can has again just been kicked down the road: in order to create pseudo-random numbers, truly random seeds are required. So, have we gained nothing?

Incorrect! Since the seed's length is of the order of $\log(n)$, and hence very short in comparison to the number of pseudo-random bits, one can enumerate and run a randomized algorithm with each of the possible seeds. For an algorithm that requires a million pseudo-random numbers, that would be no more than 20, 40, 60,… runs. Though some will not give correct answers, the majority will. And thus, we have an algorithm that requires no randomness at all; it is deterministic. The upshot is that by using a computationally hard function (if it exists), all efficient probabilistic algorithms can be transformed into deterministic counterparts.

In this, the book's next to last chapter, we explored the concept of extractors and their crucial role in harnessing randomness. We delved into their fundamental principles, such as transforming weak random sources into sequences of nearly unbiased random numbers using a short, truly random seed. We examined how extractors work, and their importance in ensuring high-quality randomness in various applications. While extractors are essential in fields directly tied to random numbers, such as cryptography and Monte Carlo simulations, their influence extends far beyond. They play significant roles in data structures, error correction, hardness of approximation, and combinatorics.

Epilogue

CHAPTER 20

The Enigma and Utility of Random Numbers

The exploration of random bits, zeroes and ones, and random decimals, zeros to nines, is not merely an academic pursuit but a journey into the heart of unpredictability and order, chance and determinism, simplicity and complexity. In this book, after discussing the fundamental nature of random numbers, we explored their varied applications and the intricate methods of producing them. As we conclude this discourse, let me reflect on the broader implications and future prospects of our understanding of random numbers.

THE ESSENCE OF RANDOM NUMBERS

At the core of our inquiry lies the fundamental question: What are random numbers? Our exploration began with an examination of their nature, from the apparent lack of pattern in sequences to the philosophical debates on whether true random numbers exist. We considered a sequence of digital or decimal numbers to be random if the members of the sequence are *u*npredictable, *u*niformly distributed, and *u*ncorrelated. (There are also distributions other than the uniform distribution, such as the normal distribution, exponential distribution, and Poisson distribution.)

DOI: 10.1201/9781003641520-27

211

We delved into parapsychology and psychokinesis, questioning the human ability to generate or predict random numbers, and ventured into the realms of entropy, complexity, and compression, measuring randomness and seeking the underlying principles that govern random numbers in natural and artificial systems.

THE UBIQUITY OF RANDOM NUMBERS

Random numbers permeate many aspects of our lives, often in ways we hardly recognize. From gaming and gambling, where they introduce an element of chance, to the more systematic applications in sortition, polling, and random sampling, random numbers serve as a foundation for fairness and representativeness. The exploration of cognitive dissonance and random walks revealed how randomness influences financial markets and decision-making processes, underscoring its ubiquitous impact on our economic and social systems.

THE ART AND SCIENCE OF PRODUCING RANDOM NUMBERS

The production of random numbers is an art, albeit steeped in scientific rigor. Moving beyond the simplicity of coins and dice, we explored mechanical and chaotic random numbers, drawing on the intricate patterns of nature. Quantum mechanics, with its inherent unpredictability, provided a perspective on the fundamental sources of randomness. These diverse methods highlight the ingenuity and creativity required to capture the essence of randomness in random numbers.

THE NECESSITY OF RANDOM NUMBERS

Random numbers are not a mere curiosity but a critical component of modern technology and science. In cryptography, they underpin the security of digital communications, safeguarding our privacy in an increasingly interconnected world. Monte Carlo methods demonstrate the power of random numbers in simulating complex scenarios and in solving difficult mathematical problems, while zero-knowledge proofs and randomized algorithms demonstrate their utility in ensuring computational efficiency and security. The fact that many randomized algorithms can be derandomized offers insights into the delicate balance between random numbers and determinism in algorithm design.

THE ILLUSION OF RANDOM NUMBERS

While true randomness is a rare and precious commodity, the art of simulating random numbers with the help of generators of pseudo-random numbers has become indispensable. These algorithms, designed to mimic the behavior of random sequences, enable practical applications even when genuine random numbers are unattainable. However, their limitations and pitfalls remind us of the ongoing challenges of generating robust and reliable series of pseudo-random numbers. Testing and validating the algorithms are essential in order to ensure their adequacy for various applications. The need for improvement in the design and implementation of random number generators never stopped and, presumably, never will.

THE FUTURE OF RANDOM NUMBERS

The interplay between random numbers and determinism, order and chaos, remains a key area of investigation. Advances in quantum computing and chaos theory may unlock new methods of generating and harnessing random numbers, while emerging applications in artificial intelligence (e.g., generative models relying on high-quality randomness), data science (e.g., Monte Carlo simulations in financial risk modeling and climate prediction), and secure communications (e.g., quantum key distribution) will drive further exploration. Interdisciplinary research combining insights from physics (studies of quantum randomness), biology (randomness in genetic algorithms and evolutionary strategies), and computer science (pseudo-random number generation for cryptography) is expected to yield novel approaches to understanding randomness and utilizing pseudo-random numbers in innovative ways.

Practical applications include – to cite just some examples – using random numbers in video game design to generate game levels, terrains, maps, or item drops; employing random numbers to distribute tasks or requests in distributed systems to avoid overload on any single server; using random numbers in blockchain consensus protocols to ensure that no single party can control the process; utilizing random numbers to generate cryptographic keys for communication networks; using random numbers in data science to build training and test sets for machine learning models; and using randomized algorithms in large-scale data processing to speed up computations that would otherwise be infeasible.

CONCLUSION

Randomness, as reflected by random numbers, highlights the complexities of our world. The study of random numbers is not just an academic endeavor but a pursuit of understanding and mastery over the unknown. Random numbers challenge our understanding, inspire creativity, and drive technological advancements. As we continue to study them, we gain a deeper appreciation for the role of chance and necessity in shaping our universe.

Endnotes

These endnotes list bibliographic references in the order in which they occur in the text. To facilitate readability, specific reference numbers have been omitted within the text itself.

PREFACE

Hasofer, A. M., "Studies in the History of Probability and Statistics, XVI. Random Mechanisms in Talmudic Literature," *Biometrika*, June 1967, Vol. 54, No. 1/2, pp. 316–321.

Talmud, tractate Baba Bathra, 122a.

Szpiro, George G., "*Numbers Rule: The Vexing Mathematics of Democracy*," Princeton University Press, 2010.

CHAPTER 1

Galton, Francis, "Dice for Statistical Experiments," *Nature*, 1890, Vol. 42, pp. 13–14.

Tippett, L. H. C., "*Random Sampling Numbers*," Cambridge University Press, 1927.

Fisher, R. A. and F. Yates, "*Statistical Tables for Biological, Agricultural and Medical Research*," Oliver and Boyd, London, 1938.

Kendall, M. G. and B. Babington-Smith, "Randomness and Random Sampling Numbers," *Journal of the Royal Statistical Society*, 1938, Vol. 101, No. 1, pp. 147–166.

RAND Corporation, "*A Million Random Digits, With 100,000 Normal Deviates*," 1955.

Brown, George W., "*History of RAND's Random Digits — Summary*," RAND Corporation, 1949. https://csrc.nist.gov/publications/detail/sp/800-22/rev-1a/final

CHAPTER 2

Goodfellow, L. D., "A Psychological Interpretation of the Results of the Zenith Radio Experiments in Telepathy," *Journal of Experimental Psychology: General*, 1992, Vol. 121, No. 2, pp. 130–144.

Einstein, Albert. *New York Times*, April 25, 1929, p. 60.

CHAPTER 3

"Helmut Schmidt: 1928–2011." *The Free Library*, 2011, Parapsychology Press, December, 17, 2021, www.thefreelibrary.com/Helmut+Schmidt%3a+1928-2011.-a0280004576

Jahn, R. G. et al., "Engineering Anomalies Research," *Journal of Scientific Exploration*, 1987, Vol. 1, No. 1, pp. 21–50.

Reichenbach, Hans, "*The Theory of Probability*," University of California Press, Berkeley, 1934/1949.

Nickerson, Raymond S., "The Production and Perception of Randomness," *Psychological Review*, 2002, Vol. 109, No. 2, pp. 330–357.

Hahn U. and P. A. Warren, "Perceptions of Randomness: Why Three Heads Are Better Than Four," *Psychological Review*, 2009, Vol. 116, pp. 454–461.

Bar Hillel, Maya and Willem Wagenaar, "The Perception of Randomness," *Advances in Applied Mathematics*, 1991, Vol. 12, pp. 428–454.

Neuberger, Allen, "Can People Behave "Randomly? The Role of Feedback," *Journal of Experimental Psychology: General*, 1986, Vol. 115, No. 1, pp. 62–75.

Green, D. R., "Testing Randomness," *Teaching Mathematics and Its Applications*, 1982, Vol. 1, pp. 95–100.

Lopes, L. L. and G. C. Oden, "Distinguishing between Random and Nonrandom Events," *Journal of Experimental Psychology: Learning, Memory, and Cognition*, 1987, Vol. 13, pp. 392–400.

Kahneman, Daniel and Amos Tversky, "Subjective Probability: A Judgment of Representativeness," *Cognitive Psychology*, 1972, Vol. 3, pp. 430–454.

Griffiths, Thomas L. and Joshua B. Tenenbaum, "*Randomness and Coincidences: Reconciling Intuition and Probability Theory*," Working paper, Stanford University, 2001.

Persaud, Navindra, "Humans Can Consciously Generate Random Number Sequences: A Possible Test for Artificial Intelligence," *Medical Hypotheses*, 2005, Vol. 65, pp. 211–214.

Figurska, Malgorzata, et al., "Humans Cannot Consciously Generate Random Numbers Sequences: Polemic Study." *Medical Hypotheses*, 2008, Vol. 70, pp. 182–185.

Chan, Ka-Shing, et al., "Random Number Generation Deficit in Early Schizophrenia," *Perceptual and Motor Skills*, 2011, Vol. 112, No. 1, pp. 91–103.

Jokar, Elham and Mohammad Mikaili, "Assessment of Human Random Number Generation for Biometric Verification," *Journal of Medical Signals and Sensors*, 2012, Vol. 2, No. 2, pp. 82–87.

Kahnemn, Daniel and Amos Tversly, "Subjective Probability: A Judgment of Representativeness," *Cognitive Psychology*, 1972, Vol. 3, pp. 430–454.

Slovic, Paul, Howard Kunreuther, and G. F. White., "Decision Processes, Rationality, and Judgments of Natural Hazards." In G. F. White (Ed.), *Natural Hazards*, 1974, Oxford University Press, New York, pp. 187–206.

CHAPTER 4

Kendall, M. G., "Studies in the History of Probability and Statistics: II. The Beginnings of a Probability Calculus," *Biometrika*, June 1956, Vol. 43, No. 1/2, pp. 1–14.

Aristotle, *Metaphysics*, 1064b, 1065a.

Kepler, Johannes, *"De nova stella in pede Serpentarii,"* 1604, as quoted in *Science et Hasard*, P. S. Coculesco, Payot (1952).

Leibniz, G. W., *"Allgemeine Untersuchungen über die Analyse der Begriffe und wahren Sätze,"* p. 289.

Clausius, R., "Ueber verschiedene für die Anwendung bequeme Formen der Hauptgleichungen der mechanischen Wärmetheorie," *Annalen der Physik*, 1865, Vol. 20, No. 7, pp. 353–400.

von Mises, Richard, "Über Zahlenfolgen, die ein Kollektiv-ähnliches Verhalten zeigen," *Mathematische Annalen*, 1933, Vol. 108, pp. 757–772.

Kolmogorov, A. N., "On Tables of Random Numbers," *Sankhya: The Indian Journal of Statistics*, Series A., 1963, Vol. 25, Part 4.

Russell, Bertrand, "Mathematical Logic as Based on the Theory of Types," *American Journal of Mathematics*, July 1908, Vol. 30, No. 3, pp. 222–262.

CHAPTER 5

"Hazard", *Encyclopedia Britannica*, 1911. https://en.wikisource.org/wiki/Page:EB1911_-_Volume_13.djvu/130

Tacitus, Germania, 24, 3–4.

Thorp, Edward E., "The invention of the first wearable computer," www.cs.virginia.edu/~evans/thorp.pdf

Gleick, James, "The Dynamical Systems Collective," *Computers in Physics*, 1988, Vol. 2, p. 40.

Bass, Thomas A. *"The Eudaemonic Pie,"* Open Road Integrated Media LLC., 2017.

Nevada Gaming Commission and the Nevada Gaming Control Board. http://gaming.nv.gov

CHAPTER 6

Gottlieb, Anthony, "Win or Lose," *The New Yorker*, July 26, 2010. Review of my *"Numbers Rule: The Vexing Mathematics of Democracy, from Plato to the Present,"* Princeton University Press, 2010. By the way, in the book I explain why majority voting is not all that it is made out to be.

For more on the St. Petersburg Paradox and related issues, see my *"Risk, Choice and Uncertainty,"* Columbia University Press, 2020.

CHAPTER 7

Kahneman, Daniel and Amos Tversk, "Subjective Probability: A Judgment of Representativeness," *Cognitive Psychology*, 1972, Vol. 3, pp. 430–454.

For the history of the Black–Scholes options pricing formula, see my book *"Pricing the Future: Finance, Physics, and the 300-year Journey to the Black–Scholes Equation,"* Basic Books, 2011.

Malkiel, Burton G., *"A Random Walk Down Wall Street,"* W. W. Norton, 1973.

Mandelbrot, Benoît and Nassim Taleb, "A focus on the exceptions that prove the rule," *Financial Times*, March 23, 2006.

CHAPTER 8

Keller, Joseph. B, "Probability of Heads," *The American Mathematical Monthly*, 1986, Vol. 93, No. 3, pp. 191–197.

Eerkens, Jelmer W. and Alex de Voogt. "Why Are Roman-Period Dice Asymmetrical? An Experimental and Quantitative Approach," *Archaeological and Anthropological Sciences*, 2022, Vol. 14, No. 134.

Diaconis, Persi and Joseph B. Keller, "Fair Dice," *The American Mathematical Monthly*, April 1989, Vol. 96, No. 4, pp. 337–339.

Kapitaniak, M., J. Strzalko, J. Grabski, and T. Kapitaniak, "The Three-Dimensional Dynamics of the Die Throw," *Chaos*, 2012, Vol. 22, p. 047504.

Trefethen, L. N. and Trefethen, L. M. "How Many Shuffles to Randomize a Deck of Cards?" *Proceedings of the Royal Society A: Mathematical, Physical and Engineering Sciences*, 2000, Vol. 456, No. 2002, pp. 2561–2568.

CHAPTER 9

Leggett, M. D., *"Subject Matter Index of Patents for Inventions Issued by United States Patent Office from 1790 to 1873,"* Government Printing Office, Washington, D.C., 1874 (Reprinted by Arno Press, 1976), Vol. II, p. 889.

Vannini v. Paine, Jun 1832, Delaware Court of Errors and Appeals. https://cite.case.law/del/1/65/

United States Patent Office, 1840. https://patentimages.storage.googleapis.com/d1/12/e4/01e3ebfdca1a6b/US1700.pdf

Richmond Enquirer, April 13, 1819.

Bellhouse, D. R., "The Genoese Lottery," *Statistical Science*, May 1991, Vol. 6, No. 2, pp. 141–148.

Patentschrift, Kaiserliches Patentamt, 1883. https://worldwide.espacenet.com/patent/search/family/034484204/publication/DE26471C?q=DE26471T

Patentschrift, Kaiserliches Patentamt, 1901. https://worldwide.espacenet.com/patent/search/family/034484381/publication/DE129735C?q=pn%3DDE129735C

Deutsches Patent- und Markenamt, 1957. https://worldwide.espacenet.com/patent/search/family/032727286/publication/DE1738167U?q=pn%3DDE1738167U

United States Patent Office, 1937. https://patents.google.com/patent/US2091883A/en

United States Patent Office, 1988. https://patents.google.com/patent/US4786056

CHAPTER 10

Ware, Willis H., "*RAND and the Information Evolution: A History in Essays and Vignettes*," RAND Corporation, 2008, p. 89,

CHAPTER 11

Einstein, Albert, Boris Podoslky, Nathan Rosen, "Can Quantum-Mechanical Description of Physical Reality Be Considered Complete," *Physical Review*, May 15, 1935, Vol. 47.

Bell, John Stewart, "On the Einstein Podolsky Rosen Paradox," *Physics Physique Физика*, 1964, Vol. 1, No. 3, pp. 195–200.

Kochen, S. and E. P. Specker, "The Problem of Hidden Variables in Quantum Mechanics," *Journal of Mathematics and Mechanics*, 1967, Vol. 17, No. 1, pp. 59–87.

Euler's Totient Theorem and Fermat's Little Theorem - Complete Proof & Intuition. www.youtube.com/watch?v=5pswKNgVZSg

CHAPTER 13

Metropolis, N., "The Beginning of the Monte Carlo Method," *Los Alamos Science Special Issue*, 1987.

Metropolis, N. and S. Ulam, "The Monte Carlo Method," *Journal of the American Statistical Association*, September 1949, Vol. 44, No. 247, pp. 335–341.

Hayes, Brian, "Orderly Randomness: Quasirandom Numbers and Quasi–Monte Carlo," Harvard Institute for Applied Computational Science, 2015-02-06, www.youtube.com/watch?v=wGflWqY41RU.

CHAPTER 14

Feldmann, Richard W. Jr., "The Cardano–Tartaglia Dispute," *The Mathematics Teacher*, 1961, Vol. 54, No. 3.

Goldwasser, Shafi, Silvio Micali, and Charles Rackoff, "The Knowledge Complexity of Interactive Proof Systems," *SIAM Journal of Computing*, February1989, Vol. 18, No. 1, pp. 186–208.

Glaser, Alexander, Boaz Barak, and Robert J. Goldston, "A Zero-Knowledge Protocol for Nuclear Warhead Verification," *Nature*, 2014, Vol. 51.

Goldreich, Oded, Silvio Micali, and Avi Wigderson, "Proofs That Yield Nothing But Their Validity or all Languages in NP Have Zero-Knowledge Proof Systems," *Journal of the Associatian for Computing Machinery*, 1991, Vol. l38, No. 1, pp. 691–697.

Cook, Stephen, "The Complexity of Theorem Proving Procedures." *Proceedings of the Third Annual ACM Symposium on Theory of Computing*, 1971, pp. 151–158.

Levin, Leonid, "Universal Search Problems." *Problems of Information Transmission*, 1973, Vol. 9, No. 3, pp. 115–116. (In Russian.)

Schwartz, J. T. "Fast Probabilistic Algorithms for Verification of Polynomial Identities," *Journal of the ACM*, 1980, Vol. 27, No. 4, pp. 701–717.

Mitzenmacher, Michael and Eli Upfal, *"Probability and Computing,"* Cambridge University Press, 2005, pp. 8–9.

Stoer, M. and F. Wagner, "A Simple Min-Cut Algorithm," *Journal of the ACM*, 1997, Vol. 44, No. 4, p. 585. However, the Karger algorithm is more efficient.

Update: Recent work shows that ultra-efficient deterministic mincut algorithms now exist, improving over Karger's randomized method. See Jason Li, "Deterministic Mincut in Almost-Linear Time" (arXiv:2106.05513) and Henzinger et al. "Deterministic Near-Linear Time Minimum Cut in Weighted Graphs" (arXiv:2401.05627) for near-linear deterministic algorithms.

CHAPTER 16

Impagliazzo, Russell, and Avi Wigderson, *"P=BPP* if *E* Requires Exponential Circuits: Derandomizing the *XOR* Lemma," *Proceedings of the Twenty-Ninth annual ACM Symposium on Theory of Computing*, 1997, Vol. 5, No. 4, pp. 220–229.

CHAPTER 17

Chaitin, G. J., "A Theory of Program Size Formally Identical to Information Theory," *Journal of the Association for the Computer Machinery*, 1975, Vol. 22, pp. 329–340.

von Neumann, John, "Various Techniques Used in Connection With Random Digits," *Journal of Research of the National Bureau of Standards, Applied Mathematics Series*, 1951, Vol. 3, pp. 36–38.

Knuth, Donald, *"The Art of Computer Programming,"* Vol. 2. 3rd edition, Addison-Wesley, 1998, p. 188.

Marsaglia, George, "Random Numbers Fall Mainly in the Plane." *Proceedings of the National Academy of Science, USA*, 1968, Vol. 61, No. 1, pp. 25–28.

Garfinkel, Simson, "Alarming Open-Source Security Holes: How a Programming Error Introduced Profound Security Vulnerabilities in Millions of Computer Systems." *MIT Technology Review*, May 20, 2008.

Goldberg, Ian and David Wagner, "Randomness in the Netscape Browser," *Dr. Dobb's Journal*, January 1996.

Jagannatam, Archana, "Mersenne Twister – A Pseudo Random Number Generator and its Variants," http://cryptography.gmu.edu/~jkaps/download. php?docid=1083

Markowsky, George, "The Sad History of Random Bits," *Journal of Cyber Security and Mobility*, January 2014, Vol. 3, No. 1, pp. 1–24.

L'Ecuyer, Pierre, "Uniform Random Number Generators". In Lovric, Miodrag (ed.). *International Encyclopedia of Statistical Science*. Springer, 2010, p. 1629.

CHAPTER 18

Ferrenberg, Alan M., D. P. Landau, Y. Joanna Wong, "Monte Carlo Simulations: Hidden Errors from "Good" Random Number Generators," *Physical Review Letters*, December 7, 1992, Vol. 69, No. 23, pp. 3382–3384.

Hayes, Brian, "The Wheel of Fortune," *American Scientist*, March–April 1993, Vol. 81.

Wolff, U., "Collective Monte Carlo Updating for Spin Systems," *Physical Review Letters*, 1989, Vol. 62, No. 4, pp. 361–364.

Grassberger, Peter, "On Correlations in "Good" Random Number Generators," *Physics Letters A*, 1993, Vol. 181, pp. 43–46.

Bauke, Haiko and Stephan Mertens, "Pseudo Random Coins Show More Heads Than Tails," *Journal of Statistical Physics*, 2004, Vol. 114, pp. 1149–1169.

James, Frederick, and Moneta, Lorenzo, "Review of High-Quality Random Number Generators," *Computing and Software for Bio Science*, 2020, Vol. 4, No. 2.

Segert, Simon, "A proof that HT is more likely to outnumber HH than vice versa in a sequence of n coin flips," https://arxiv.org/abs/2405.16660.

CHAPTER 19

Samuelson, Paul A., "Constructing an Unbiased Random Sequence," *Journal of the American Statistical Association*, 1968, Vol. 63, No. 324, pp. 1526–1527.

Blum, M., "Independent Unbiased Coin Flips from a Correlated Biased Source: A Finite State Markov Chain," *Foundations of Computer Science, Annual IEEE Symposium*, 1984, pp. 425–433.

Stein, P. R. and S. M Ulam, Rozpr. Mat. 39.. Reprinted in: Stanislaw M. Ulam, "*Analogies between Analogies: The Mathematical Reports of S. M. Ulam and His Los Alamos Collaborators,*" University of California Press, 1964, pp. 349–350.

Ulam, S. et al., *"Analogies Between Analogies: The Mathematical Reports of S.M. Ulam and His Los Alamos Collaborators,"* University of California Press, 1990.

Collins, J. J., M. Fanciulli, R. G. Hohlfeld, et al., "A Random Number Generator Based on the Logit Transform of the Logistic Variable," *Computers in Physics*, 1992, Vol. 6, p. 630.

Chor, B. and O. Goldreich, "Unbiased Bits from Sources of Weak Randomness and Probabilistic Communication Complexity," *SIAM Journal on Computing*, April 1988, Vol. 17, pp. 230–261.

Santha, Miklos and Umesh Vazirani, "Generating Quasi-random Sequences from Semi-random Sources," *Journal of Computer and System Sciences*, 1986, Vol. 33, p. 81.

Manindra Agrawal, Neeraj Kayal, and Nitin Saxena, "PRIMES is in P," *Annals of Mathematics*, 2004, Vol. 160, pp. 781–793.

CHAPTER 20

For an advanced text on everything to do with random numbers – and much, much more – see Wigderson, Avi, *"Mathematics and Computation: A Theory Revolutionizing Technology and Science,"* Princeton University Press, 2019.

Free download at: www.math.ias.edu/files/Book-online-Aug0619.pdf

Index

For Product Safety Concerns and Information please contact our EU
representative GPSR@taylorandfrancis.com
Taylor & Francis Verlag GmbH, Kaufingerstraße 24, 80331 München, Germany

www.ingramcontent.com/pod-product-compliance
Lightning Source LLC
Chambersburg PA
CBHW060830170526
45158CB00001B/124